Contents

managing environments
IN BRITAIN
& IRELAND

JOHN CHAFFEY

Hodder & Stoughton
A MEMBER OF THE HODDER HEADLINE GROUP

To Zoë, who will grow up in a different Britain to the one pictured within

ACKNOWLEDGEMENTS

Research for this book has taken me to many parts of the United Kingdom and the Irish Republic. The endless variety which these islands have to offer has been a constant reward for the many days of travel. Everywhere, people were willing to talk to me about the environments whose management was their responsibility. From the city halls of Edinburgh and Belfast to the visitor centres of Glencoe and Once Brewed, managers provided me with new ideas and valued resources. My thanks go to Mike Amphlett, Neil Allsop and Geoff Davies of West Dorset District Council; Michael Roberts and Jeremy Fitch of the Northern Ireland Development Board; Bob Huggins and Dave Cooling of the Environment Agency; Robin Davies of the Forestry Commission; Barry Entwhistle of Blackburn Borough Council; Mike Finnemore of English Nature; Ian Forshaw and Adrian McLaughlin of Forest Enterprise; Andy Hill of English China Clay; Nuala McKeagney of Belfast City Council; Tony McKenna of Bord na Mona; Andy Megan of the Dorset Heritage Coast; Diane Milne of Dundee City Council; Lewis Porter of Craigavon Borough Council; Peter Stone of Camas Aggregates; Peter Strong of Edinburgh City Council; David Venner of the South West Path Project; Derrick Warner of the National Trust, Glencoe; Andrew Whittaker of South West Water; and Sally York of the Forestry Commission. Margaret and John Priestley provided us with an essential base in Edinburgh. Julia Morris and Katherine Stathers at Hodders have been sympathetic and helpful editorial colleagues. Finally, as always, my thanks go to my wife, Ruth, for accompanying me on the many long journeys, and for her constant and unfailing support and encouragement.

The author and publishers would like to thank the following for permission to reproduce copyright material and photos in this book:
BAA Heathrow, Figure 9.5.2; Corbis, Figure 10.4.1; Crown Copyright, Figure 10.3.2; The Guardian, Figures 6.3.15, 7.6.2, 8.1.3 and 8.4.1; The Independent, Figures 3.4.4, 6.1.1, 7.3.1, 7.3.6 and 9.3.4; The Independent on Sunday, Figure 3.1.2; The Irish Times, Figure 4.4.8; The Observer, Figure 8.1.2; Purbeck Advertiser, Figure 4.3.6; Swanage and Wareham Advertiser, Figure 4.3.1.

All other photos belong to the author.

A catalogue record for this title is available from The British Library

ISBN 0 340 65559 3

First published 1997
Impression number   10 9 8 7 6 5 4 3 2 1
Year   2002 2001 2000 1999 1998 1997

Copyright © 1997 John Chaffey

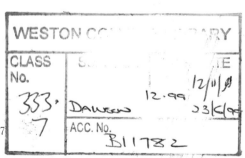
Typeset by Fakenham Photosetting Limited, Fakenham, Norfolk NR21 8NL
Printed in Hong Kong for Hodder & Stoughton Educational, a division of Hodder Headline Plc, 338 Euston Road, London NW1 3BH by Colorcraft Limited.

Introduction

In the late twentieth century we have become increasingly conscious of the environment around us. Environmentalism, as the quotation suggests, is very much in vogue, but we have to be cautious in assessing its role in the development of management strategies. This book sees the environment as consisting of two interdependent yet distinct components, the natural, and the human. The natural environment is the physical and living world in which we find ourselves, and the human environment is the one which we have created within the opportunities and constraints that the physical environment offers us.

For these environments to survive as healthy systems, they have to be managed. The natural environment, if unmanaged, will not survive the increasing pressures to which it is submitted. Increasingly, as we understand more about the functioning of the natural environments, it is clear that its management, in order to be successful, has to operate within the inherent processes of its systems. Experience is now beginning to show us that if management is too rigid, and does not recognise the need to maintain the equilibrium that exists within natural processes, then the whole system becomes unbalanced. This applies equally well to physical systems, such as coasts or rivers, and to living ecosystems.

Human environments require equally careful management. They have necessarily to co-exist with natural systems, and maintaining the correct balance is difficult. Urban, industrial and transport systems, unless managed properly, can damage the functioning of natural systems. Such systems need to function efficiently within themselves too. For instance, if we lose control of processes within the urban system, then sprawl and all of its consequences follow. Understanding the processes that operate within human systems is equally as important as comprehending those that function within natural ones. Management of human environments thus has a dual role: to monitor and adjust its interaction with the physical environment, and to regulate processes within the human environment itself to ensure its healthy functioning.

This book therefore falls into two sections. Part One considers the management of natural environments. Coasts, rivers, forests and woodlands are all natural systems. The upland environment involves a number of natural systems, which require sensitive management in order to retain a fragile and delicate equilibrium. Physical resources, such as minerals and water, are part of the natural environment, but their exploitation involves complex interaction with human systems. Part Two examines the way in which human environments are managed. Since the population of Britain is so overwhelmingly urban, and the position of Belfast and Dublin is so dominant in Northern Ireland and the Irish Republic, urban environments are considered first. Industrial environments, still suffering the consequences of de-industrialisation in different parts of the United Kingdom are examined next. To balance the theme of adjustment to de-industrialisation, East Anglia and the west of Ireland are studied as areas where potential for technological and service industries exists and is being developed. Rural environments need special management skills to maintain a social fabric now much under threat from the twin pressures of counterurbanisation and service rationalisation. Managing transport systems embraces all three of these human environments, and increasingly involves monitoring the effect of transport on the natural environment, and devising suitable and effective controls on emissions. A review of the leisure and tourism environments completes the human section, emphasising the importance of achieving a balance between service provision and environmental conservation. The examination of each environment begins with a brief statement of the key ideas involved within the chapter. A general section looks at management structure and management issues in each

> *Environmentalism is currently the flavour of the month. It will not go away, but it cannot afford to push its luck.*
>
> Tim O'Riordan in The Environment in 'Policy and Change in Thatcher's Britain.'

particular environment in Britain and
Ireland. Case studies are selected to give a
balanced overview of Britain and Ireland.
Longer detailed case studies such as those
of the Dorset Coast, the New Forest,
Glencoe and Dublin are supported by
shorter examples to give a full overview of
management issues and problems in Britain
and Ireland. The location of the case
studies is shown on the map inside the
front cover.

MANAGING NATURAL ENVIRONMENTS

1

Coastal Management

KEY IDEAS

- Coastal management embraces the complexities of both the physical and the human environment
- Shoreline management plans based on coastal cells appear to create a new way forward in coastal management
- Incomplete understanding of physical shoreline processes can often cause management schemes to fail; natural processes of defence and protection should be an integral part of any management plan
- Human occupancy of the shoreline often disturbs and distorts shoreline processes
- Coastal defence works should always be designed in a sympathetic manner, and include the retention of a natural beach
- Land-use planning in the coastal zone should consider carefully the vulnerable areas subject to future sea-level rise and increasing storminess

1.1 Introduction: the Background to Coastal Management

Coastal zone management involves a complex and bewildering interplay of many different elements of the physical and human environment. Success in managing the physical environments depends firstly on achieving a thorough understanding of the processes that operate along the coast, and secondly on deciding what adjustments, if any, are necessary in order that people may live and work harmoniously within the complexities of the coastal environment.

Coastal management in Britain and Ireland has not really benefited from an integrated approach from the authorities responsible for management. Legislation has failed to treat the problems of the coast as a whole,

and there is still a certain ambiguity concerning responsibilities for management. There is so much variety in the physical form of the coasts of Britain and Ireland, and in the level and nature of its human occupancy, that management is often fragmented and unco-ordinated. Much of the current management of coastal zones operates under various designation orders that tend to concentrate on conservation and amenity questions (Figure 1.1.1). Many designations put an emphasis on planning procedures, such as those concerned with National Parks (where they contain stretches of coastline), and Areas of Outstanding Natural Beauty (AONBs) in England and Wales. In

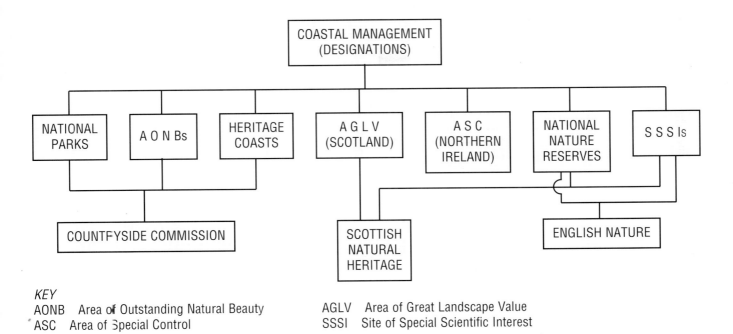

KEY
AONB   Area of Outstanding Natural Beauty
ASC   Area of Special Control

AGLV   Area of Great Landscape Value
SSSI   Site of Special Scientific Interest

Figure 1.1.1 *Coastal Management: Designations*

Scotland, Areas of Great Landscape Value (AGLVs) are recognised, and in Northern Ireland the Area of Special Control includes all of the scenic coast of Northern Ireland. The designation of Heritage Coasts on England and Wales (Figure 1.1.2) has been a response by the Countryside Commission to various initiatives of the 1960s.

In Scotland, 'Coastal Planning Guidelines' was produced by the Scottish Office in 1974 in response to the rapid growth of on-shore support facilities for the oil and gas industry. The Scottish Countryside Commission undertook a series of studies on the Scottish Coast in the 1970s and 1980s. Its role in coastal management has been largely taken over by Scottish Natural Heritage (SNH). SNH has completed work that has recognised the existence of coastal

Figure 1.1.2 *Hartland Quay Devon*

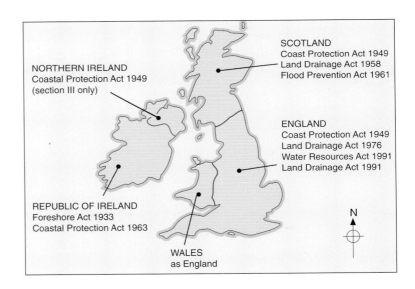

NORTHERN IRELAND
Coastal Protection Act 1949
(section III only)

SCOTLAND
Coast Protection Act 1949
Land Drainage Act 1958
Flood Prevention Act 1961

ENGLAND
Coast Protection Act 1949
Land Drainage Act 1976
Water Resources Act 1991
Land Drainage Act 1991

REPUBLIC OF IRELAND
Foreshore Act 1933
Coastal Protection Act 1963

WALES
as England

N

Figure 1.1.3 *Legislation: Coastal Management*

circulation cells, and is currently instituting work that will lead to sustainable management plans. In Northern Ireland most of the management is in the hands of non Government Organisations, such as the National Trust. Coastal management in the Republic of Ireland not only suffers from a lack of clear guidelines, but also from chronic underfunding.

Coastal erosion and shoreline protection in Britain and Ireland are managed through a range of legislative measures (see Figure 1.1.3). In the United Kingdom a

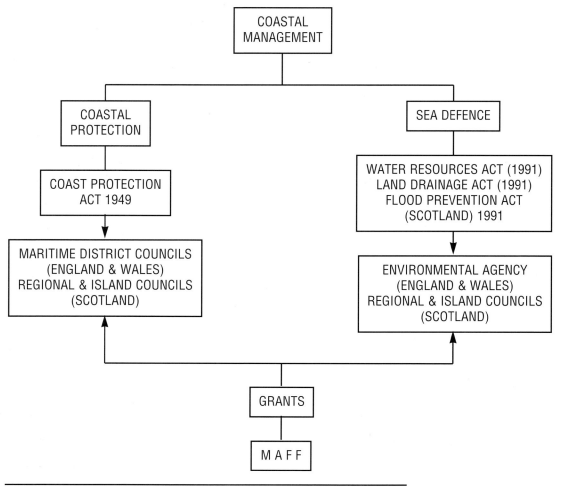

Figure 1.1.4   *Legal and administrative framework: Coastal protection and sea defence*

distinction exists between coast protection and sea defence. A Department of Environment Review stresses the difference between the two: coast protection is the protection of land from erosion and encroachment by the sea: sea defence is the defence against flooding and inundation. The Coastal Protection Act of 1949 is operated through a range of maritime authorities, who carry out erosion control work, usually grant aided by MAFF (Ministry of Agriculture, Fisheries and Food).

The Water Resources Act of 1991, and the Land Drainage Act also of 1991, deal with sea defence. The Environment Agency has responsibility for the carrying out of coast defence works, except where defences are privately or local authority owned. MAFF is similarly responsible for the awarding of grants for work under the 1991 Acts. The legal and administrative framework for coastal defence is shown in Figure 1.1.4.

An important recent development in coastal managment along the central southern coast of England has been the creation of SCOPAC (Standing Conference on Problems Associated with the Coastline). Similar groups have been formed to give a coverage of the whole of the coastline of England and Wales. SCOPAC has successfully followed a two-fold policy approach. It has developed a number of important research initiatives investigating process and change along the coast, and secondly, it has been active in persuading the Government of the need for a more integrated approach to coastal management. The main outcome of the latter has been the initiation of work on Shoreline Management Plans.

SHORELINE MANAGEMENT PLANS

A Shoreline Management Plan is defined as 'a document which sets out a strategy for

Figure 1.1.5 *Boundaries of major coastal cells*

interruption to this movement should not have any noticeable effect on neighbouring sediment cells. It would appear that the sediment cells form natural units for the development of Shoreline Management Plans, although in the first instance it is likely that sub-cells (Figure 1.1.6) will form the initial unit for the development of the Plans. On a smaller scale Management Units within the Plan will probably form a first basis for appraisal (Figure 1.1.6). Shoreline Management Plans aim at working with nature, and not against it, embodying the Government guidelines of 1993. Actively encouraged by the Government, Shoreline Management Plans are essentially voluntary, rather than statutory, but they should have a powerful influence on the way in which the coastline is managed in the future.

coastal defence for a specified length of coast taking account of natural coastal processes and human and other environmental influences and needs'. The concept of these plans is a management response to recent research that suggests that the coastline of England, Wales and Scotland appears to be divided into 20 major sediment cells (Figure 1.1.5). Within each sediment cell movement of sand and shingle is relatively self-contained, and any

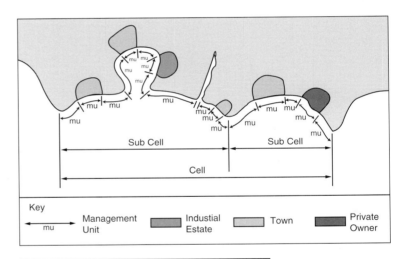

Figure 1.1.6 *Sediment cells and Management Units*

1.2 Case Studies of Coastal Management

Two case studies are chosen to illustrate the challenges presented to coastal management in Britain and Ireland. The Dorset Coast embraces a whole range of physical and human environments. Within about 140 km, the coast shows enormous physical variety. Broad embayments have been eroded in softer materials to form Lyme Bay in the west, Weymouth Bay in

the centre and Christchurch and Poole Bays in the east. Much of the coast is cliffed, but vivid contrasts exist between the West Dorset cliffs continually threatened by landslips, and the vertical cliffs in the tough limestones along the south coast of the Isle of Purbeck and between the scenically dominant chalk cliffs of East Dorset, and the totally managed cliffs of Bournemouth. In human terms there is a marked contrast between the urban-fringed coastline of Bournemouth and Poole, the south coast of the Isle of Purbeck where coastal settlements are few, the long empty stretch of Chesil Beach, and the small coastal villages and towns of West Dorset.

The north east coast of Ireland extends from the mouth of the River Foyle at Londonderry to Larne. Considerably longer (227 km) than the Dorset coast, it again shows a great variety in its coastal landforms. As in Dorset, much of the coast is fringed with cliffs, although the basalts that outcrop along much of the coast between the Giant's Causeway and Larne form more rugged scenery than is found in Dorset. At the western end of this coastline there are important depositional landforms at Magilligan and on either side of the Bann estuary at Castlerock and

Portstewart. Interesting comparisons may be established between these dune complexes and those of the South Haven Peninsula, at the southern entrance to Poole Harbour.

No settlement the size of Bournemouth and Poole occurs on the north east coast of Ireland. Significant sections of the coast, such as the stretches between the Giant's Causeway and Ballycastle, and Ballycastle and Cushendun carry little settlement. Resorts such as Portstewart, Portrush and Ballycastle take advantage of favoured situations along the north coast, and small villages such as Cushendun, Cushendall and Carnlough occupy sheltered positions at the heads of small bays on the east coast. Although this north east coast of Ireland is perhaps less accessible to tourists than Dorset, the Giant's Causeway remains probably the most visited natural site in Ireland.

1.3 Managing the Coast of Dorset

THE GEOLOGY OF THE DORSET COAST

Figure 1.3.1 shows the geology and structure of the Dorset Coast. The rocks exposed along the coast are of Mesozoic and Tertiary age, becoming progressively younger towards the east. In west Dorset Jurassic clays, and sands are overlain by permeable Cretaceous sands and

sandstones (Figure 1.3.2) and this has created the necessary conditions for extensive landslipping along the coast between West Bay and Lyme Regis. Jurassic rocks outcrop again between Weymouth and the headland of White Nothe and their disposition again encourages landslipping. Eastwards from White Nothe as far as Handfast Point,

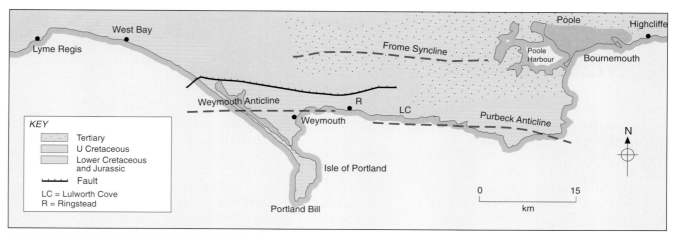

Figure 1.3.1 *Geology and Structure of the Dorset coast*

Figure 1.3.2   *Black Ven*

Figure 1.3.3   *Durdle Door and Swyre Head*

Jurassic and Cretaceous rocks combine to produce some of the most dramatic scenery on the south coast of England (Figure 1.3.3). From the mouth of Poole Harbour eastwards unresistant sands, clays and gravels of Tertiary age combine to produce unstable cliffs, which have required expensive management schemes along a coastline that is densely populated compared with the rest of Dorset.

SEDIMENT CELLS

A glance at Figure 1.1.5 shows that the Dorset coast lies within sediment cells five (Portland Bill to Land's End) and six (Portland Bill to Selsey Bill) as identified on the map. Figure 1.3.5 shows the sub-cells that are in operation along the Dorset coast, i.e. Lyme Bay, Weymouth Bay and Purbeck, and Poole and Christchurch Bays. In the Lyme Bay cell two distinct sections appear to exist. Chesil Beach appears to be a closed sediment circulation with no contemporary supply. The shingle structure appears to have been formed when gravel deposits in Lyme Bay (at that time dry)

were swept landwards by a post-glacial rising sea level 14 000–7000 years BP (before present)). Between 5000–4000 years BP this supply was cut off, and the landward movement of Chesil Beach slowed considerably. In the west of the Lyme Bay cell the situation is different, with very considerable inputs (Figure 1.3.4). Beaches are relatively small, indicating a high level of redistribution of sediment, much of it to offshore gravel banks.

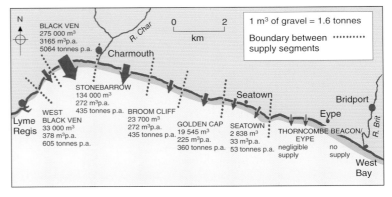

Figure 1.3.4   *Sediment supply, Lyme Bay*

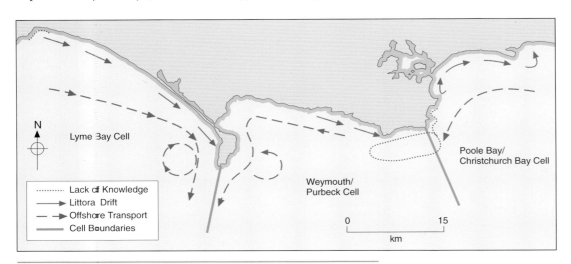

Figure 1.3.5   *Sediment sub-cells and sediment pathways on the Dorset Coast*

The mode of operation of the Weymouth Bay, and Purbeck cell is less well understood. Along a coastline where beaches are separated by headlands, there appears to be little opportunity for longshore movement of material, and shore-normal movements appear to be blocked often by submarine block barriers. More complex changes have occurred in the Weymouth Bay section of the cell, and these appear to be related to the construction of Portland Harbour in the 1850s. Sediment inputs now appear to be concentrated just to the east of Weymouth at Bowleaze Cove, particularly at Furzy Cliff. The most easterly cell is in Poole and Christchurch Bays. Little evidence exists of longshore drift in Swanage Bay, but there may be some south-north drift in Durlston Bay to the south. Sediment inputs in Poole Bay have been much reduced since the major coastal defence structures were built at Bournemouth in the 1890s. Some input comes from the eroding cliffs at Hengistbury Head at the eastern end of Poole Bay, but this was estimated at only 4000 m^3 per annum in 1985. Important defence works at Highcliffe in the east of the cell have also reduced sediment inputs. Beach levels appear to have fallen over the last 80 years, and beach nourishment has been necessary to maintain the beach resource for tourism. Figure 1.3.7 shows the current dynamic status of the Dorset Coast.

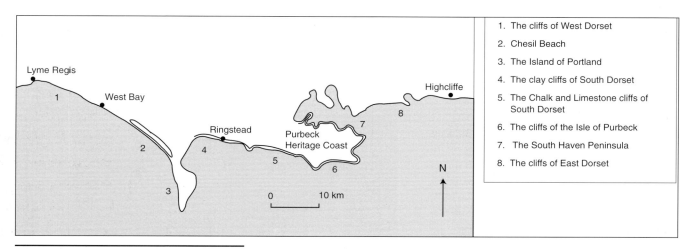

Figure 1.3.6 *Dorset Coast: Landform divisions*

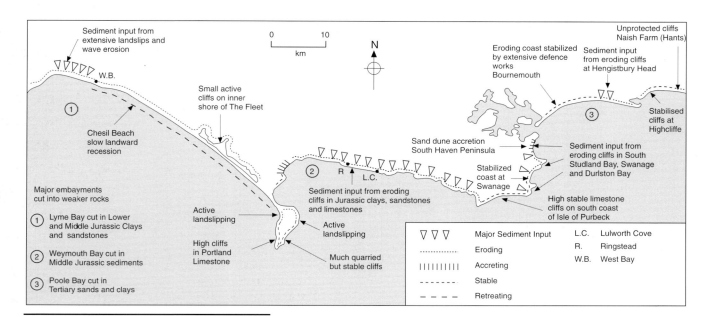

Figure 1.3.7 *Dorset Coast: current dynamic status*

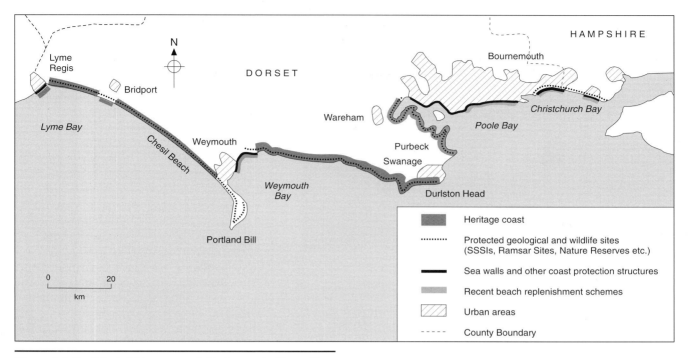

Figure 1.3.8   *Principal Management Initiatives on the Dorset Coast*

1 Compare and contrast the patterns of sediment movement in the three sub-cells of the Dorset Coast.
2 What evidence is there that human interference has modified the workings of the cells?

SOME MANAGEMENT ISSUES ON THE DORSET COAST

From the preceding sections it will be seen that there is a whole range of challenges to coastal managers along the Dorset Coast. It will have been noted that many of the cliffs are unstable in both West and East Dorset, and require a management response. Other physical problems, such as coastal flooding at Chesilton and West Bay also need remedial action. Furthermore, coastal management inevitably also extends to other fields. With such large numbers of visitors, management of recreational resources has also to consider conservation priorities in order to establish an acceptable balance between the two. Control of pollution and maintenance of water quality are two related management requirements. Figure 1.3.8 shows some management initiatives on the Dorset Coast. A number of case studies are now examined to show the range of management initiatives operating along the Dorset Coast. The map (Figure 1.3.1) shows the location of the following:

● Management of recreation and conservation: the Purbeck Heritage Coast (with special reference to the Lulworth Ranges and West Lulworth)
● Coastal protection at a sensitive geological site: Ringstead
● Cliff stabilisation, flood protection and harbour entrance management: West Bay

A CONFLICT OF INTERESTS: THE PURBECK HERITAGE COAST

The concept of Heritage Coasts was first established in 1970, with the twin aims of conservation and facilitating enjoyment. The aims of Heritage Coasts were restated in 1991. Although they remained broadly the same, the revision also embraced the management of inshore waters, and sustainable social and economic developments.

Why is Purbeck a Heritage Coast? 'Because it is special and merits special care. With its spectacular scenery, world famous geology and superb wildlife, it is both diverse and delightful. It is also home, workplace or source of relaxation for many thousands of people. Inevitably conflicts will arise, but . . . understand the need for care, thought and ingenuity in our everyday actions if this rich heritage is to be enjoyed by tomorrow's children.'

Dorset County Council leaflet: The Purbeck Heritage Coast, celebrating 20 years of successful co-operation and the renewal of the European Diploma for Conservation.

Figure 1.3.9   *Arish Mell, Lulworth Gunnery Ranges*

much valued ecosystems, geological sites of world importance and coastal scenery of the highest order. The coast attracts millions of tourists every year, and the apparent conflict is further complicated by the existence of the Lulworth Gunnery Ranges in the central stretch of the southern section (Figure 1.3.9). The Ranges, now without any permanent population since 1943, lie in juxtaposition with Lulworth Cove, one of the great 'honey-pot' sites of the Heritage Coast (Figure 1.3.11). It is proposed to examine the contrasts that exist between these two adjacent stretches of coastline, and the management problems they present.

The map (Figure 1.3.12) shows the extent of the Lulworth Ranges. This section of the coast is particularly rich geologically, with a whole series of type sections, and includes the famous Fossil Forest just to the east of Lulworth Cove. Coastal scenery embraces spectacular cliffs cut in Chalk and Jurassic limestones with more sheltered bays eroded in the softer Wealden Beds (see Figure 1.3.11). Flora and fauna carry equal distinction. Much of the unimproved limestone grassland in Dorset lies within the Ranges, and a number of rare species of plants such as Carrot Broomrape grow on

The extent of the Purbeck Heritage coast is shown on the map, Figure 1.3.10. It embraces two stretches of coastline. In the north it includes all of the coast of Poole Harbour outside of the urban and industrial zone of Poole and Hamworthy, and the coast as far south as the northern outskirts of Swanage. The southern stretch runs from Peveril Point in Swanage to Redcliff Point overlooking Bowleaze Cove to the east of Weymouth. Both sections possess

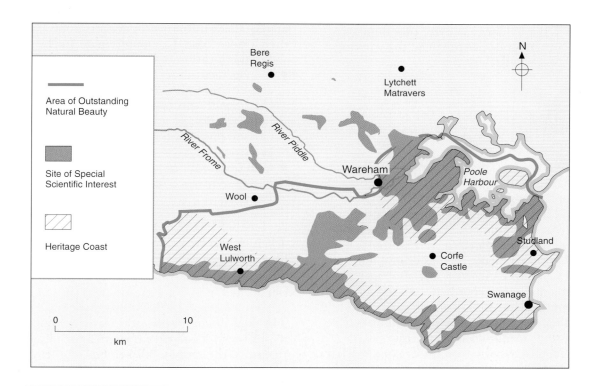

Figure 1.3.10   *The Purbeck Heritage Coast*

the cliffs, which also possess the strongest national population of wild cabbage. Species of butterflies include the distinctive Adonis Blue and the Lulworth Skipper.

The ranges are rich in archaeology too, with the important Iron Age forts at Bindon Hill and Flower's Barrow. Until the military acquisition there were important farming communities in the Ranges, one of the most important being that based on Tyneham in its own remote valley, leading down to Worbarrow Bay. Management of the Lulworth Ranges is in the hands of the Ministry of Defence. Overall responsibility lies with the Range Officer at Lulworth Camp, but day-to-day work in the field is carried out by a team of eight wardens. Apart from military training, which is the main reason for the existence of the Ranges, management involves conservation, coastal management, and public access. Conservation matters involve a great deal of liaison with official and voluntary bodies. Routine discussion of conservation matters is combined with the commissioning of periodic reports on issues such as the flora of Chalk grassland. Restricted public access and a well organised, imaginative conservation programme are now seen as being extremely helpful in safeguarding the ecological status of the Ranges.

Coastal management, within the Ranges, is

Figure 1.3.11   *Lulworth Cove, Dorset*

seen as maintaining safety and security along the Coastal Path, and in coastal waters. Weekly and annual inspections of the coastal path ensure that it is safe for walkers. Where coastal erosion endangers the path, remedial measures are taken, and, from time to time, sections of the path are diverted in order to allow them to recuperate after heavy usage has caused damage. Recreational activity, and fishing in local waters is monitored so that it remains in harmony with firing schedules. Public access to the Lulworth Ranges has long been controversial. The Range walks were established after the Nugent Report of 1974. Access is through a series of gates

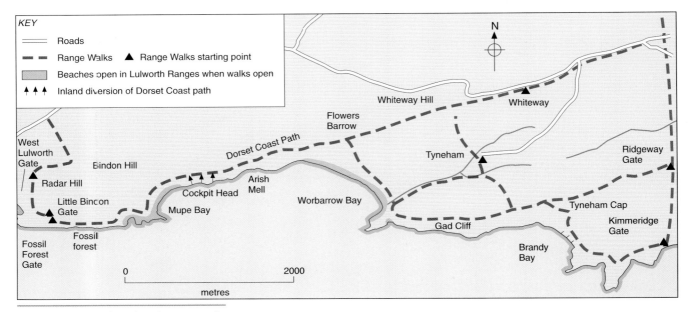

Figure 1.3.12   *The Lulworth Range Walks*

(see Figure 1.3.12). In practice the Walks are open every weekend, (apart from those scheduled for firing) and for longer periods at Easter, the Spring Bank Holiday and in the school holidays in summer.

It is now asserted by many that the military presence in the Lulworth Ranges is broadly beneficial from the long term ecological viewpoint.

Lulworth Cove (Fig 1.3.11) and the stretch of coast as far as Durdle Door are one of the most intensively used sections of the Dorset Coast. The main recreational resources and provision are shown on the map, (Figure 1.3.13). Difficulties arise in the management of this section of the Heritage Coast, simply because different groups share responsibility. Dorset County Council, through its Heritage Coast officials, has responsibility for conservation of flora, fauna, geological and archaeological sites, and for monitoring of recreational activities to ensure compatibility with conservation aims. Monitoring of the condition of the Dorset Coast Path, and beach access paths at Durdle Door, together with taking appropriate remedial action, is another important responsibility. Visitor management, and the provision of visitor facilities is the accepted responsibility of the local landowners, the Weld Estates. Three main areas of management appear to give cause for concern at the moment:

1. Car Parking. At present the main car park (Figure 1.3.14) at Lulworth is operated by the Weld Estates. It is gravel

Figure 1.3.14 Car park at Lulworth Cove, August 1996

surfaced, and has overflow capacity in adjacent fields. The average length of stay is 1 hour 43 minutes, and it is a considered view that this relatively short stay does not bring enough returns from visitors for the local economy. It follows that better visitor facilities at Lulworth would encourage longer stays. The new Heritage Centre, adjacent to the Car Park, was opened in 1994. Car Park visitors are channelled into the Centre (entrance is part of the parking fee). At the moment the Car Park occupies a very prominent position, and is regarded as being detrimental to the overall landscape at Lulworth. Pressure to relocate exists, but the search for a suitable alternative sites has as yet proved unsuccessful.

2. Durdle Door Camping and Caravan Site. This visitor facility has been operated by the Weld Estates for over 70 years, and occupies a prominent position at the top of

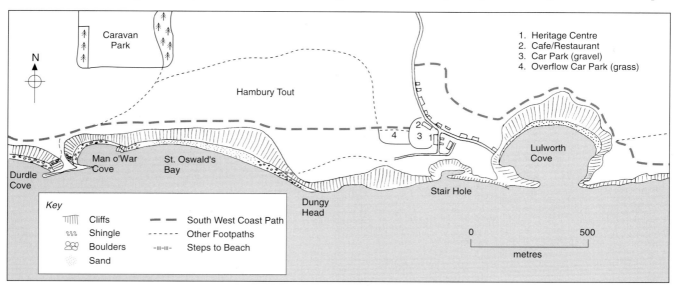

Figure 1.3.13 *Lulworth Cove and Durdle Door, main recreational resources*

the cliffs overlooking Durdle Door. In a relatively exposed site, it can also be regarded as a landscape intrusion. The Heritage Coast authority has encouraged the planting of trees along the east and west perimeters, but longer term landscaping proposals are awaited. Alternative sites have been sought, without any firm schemes being considered.
3. The erosion of the Coastal Path between Lulworth and Durdle Door is a matter of considerable concern. It is probably the most intensively used section of the South West Way, with over 200 000 walkers using it per annum. Erosion levels are high, with depths of up to 30 cm occurring in one year. In the last 15 years the path has increased in width by 2 m. Erosion has caused the whole of the adjacent hillside to begin to slip (Figure 1.3.15). Remedial measures have been attempted, with little success so far, and the whole of the path is now in a dangerous state, with broken ankles a common occurrence in summer.
Paths leading down to Man o'War Cove and Durdle Cove are under similar pressure. New two metre wide steps have been installed to encourage two way flows, and large sections of the Durdle Promontory have been fenced off for reseeding. One of the main difficulties at Durdle Door is that the nearby cliffs of the Wealden Beds are highly unstable and visitor pressure has exacerbated an already difficult situation.

STUDENT ACTIVITY

Lulworth Cove clearly requires an integrated management plan.
1 Suggest what should be the main terms of reference for such a management plan.
2 Using a blank map of the site, annotate it with suggestions for
  a additional visitor facilities
  b alternative proposals for the diversion of the Coastal Path from Lulworth Cove to Durdle Door
  c location of interpretive facilities away from the main car park.
3 Compare the management of the Lulworth – Durdle Door honeypot with that of the Lulworth Ranges. Are there any lessons to be learnt?

Figure 1.3.15   *Eroded footpath from Lulworth Cove to Durdle Door*

COASTAL PROTECTION AT A SENSITIVE GEOLOGICAL SITE: RINGSTEAD BAY

The Problem
Ringstead Bay lies some 15 km to the east of Weymouth. The coast is designated a Site of Special Scientific Interest, largely because of its geological importance. British sections for the boundary of the Oxfordian and the Kimmeridgian, together with the geologically important Ringstead Waxy Clays, and the Ringstead Coral Bed are exposed in the cliffs and along the foreshore. They have been studied for nearly 170 years, and are regarded as a classic section for the study of the rocks of the Jurassic Period.
Above the low cliffs are a number of residential properties and a busy caravan site. Shingle protects the cliffs from erosion, but much has been eroded by storms in 1989 and late 1990. Clay exposed along the foreshore has been readily eroded, and this has led to further erosion in the clay cliffs behind. Erosion of the soft material in the cliffs has led to slipping along much of the shore in front of the properties and the caravan site. Further difficulties arose because of heavy rainfall in 1994. Septic tanks in the residential properties have become overloaded, leading to sewage pollution along the beach. Saturation of the clays on which the houses have been built has led to slumping, and the main stream flowing

into the bay has eroded strongly enough to cut a deep gully back from the cliffs.

The solution (Figure 1.3.17)
Shingle loss from the central beach of Ringstead Bay does not seem capable of being replenished naturally. During the storms of 1989–1990 shingle seems to have been removed offshore by **rip currents** so that it is lost to longshore movement for a long time. In the bay itself there appears to be no clear direction of longshore drift, but the loss of shingle offshore has led to falling beach levels both to the east and the west of the central area from which it was removed during the storms.
Beach replenishment, using material brought from the nearby Warmwell gravel pits, will nourish the beach to an extent that the **berm** built up will slope out at 1 in 5 to the same distance as the beach in the 1980s. Wave action will modify this to the original, pre-1989 levels of the beach. However, this replenishment will cover the valued exposures of the Ringstead Waxy Clays, and the Ringstead Coral Bed, but leaves the western exposure of the Coral Bed untouched.
Rock armour will be used to protect the caravan park on the seaward side, together with a groyne at the eastern end of the caravan park (Figure 1.3.16). Further rock armour will protect the Kimmeridge Clay Cliffs to the east of the groyne. Inevitably

Figure 1.3.16   Coastal defence works at Ringstead Bay

other important exposures will be lost by being covered with this rock armour.
On the landward side an attempt has been made to improve the drainage problems that exacerbated the difficulties. Three concrete culverts have been provided to drain off water to the beach.

The controversy
Put simply the conflict is between protection of property and loss of a much valued geological site of world importance. Although the dispute did not extend to the level of a Public Enquiry, local geologists, much concerned by the proposals, did enlist the assistance of professional geologists. Site meetings with West Dorset District Council followed, where a full discussion of the geological implications of the protection work took place.

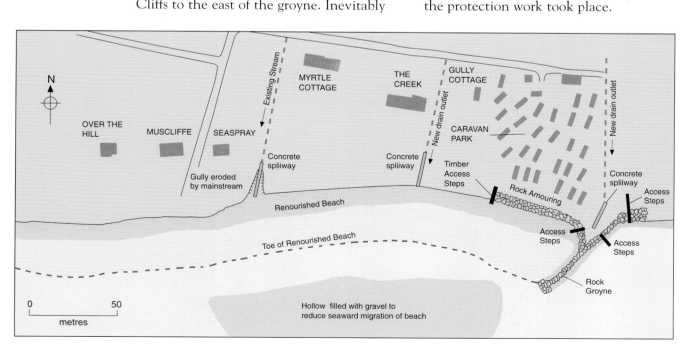

Figure 1.3.17   Ringstead Coastal Protection scheme

In the event the scheme went ahead, and is now complete. Whether it will be more successful than previous schemes, only time will tell.

STUDENT   1.3   ACTIVITY

ROLE PLAY EXERCISE

There appears to be a number of key players in this controversy.
1 Local amateur geologist, used to collecting fossils from this site.
2 Professional geologist, who has used the site for research over a period of years.
3 Caravan owner on the site immediately threatened by the loss of beach shingle.
4 Owner of residential property, The Creek.
5 Representative of Engineering consultants who prepared the report on the coastal protection problem.
6 Representative of West Dorset District council, who are the clients for whom the work will be done.
Hold a class discussion on the controversy, with each role being taken by a different member. A Chairman should be appointed to oversee the proceedings. At the end a vote should be taken on whether the scheme should have gone ahead or not.

CLIFF STABILISATION, FLOOD PROTECTION AND HARBOUR ENTRANCE MANAGEMENT, WEST BAY

The problems at West Bay are altogether more complex than in the other case studies so far reviewed. West Bay is a small coastal settlement (Figure 1.3.18) on the West Dorset Coast, distinguished by its two parallel harbour walls, through which the River Brit makes its exit into Lyme Bay. To the west of the harbour walls the cliff beyond the Esplanade has long been a problem for coastal managers on account of its instability. Both the eastern and western parts of West Bay are prone to flooding when the sea defences on the west, and the shingle bank on the east (effectively the end of Chesil Beach) are overtopped. The harbour walls, both listed structures, are becoming increasingly vulnerable, and an overall scheme for their improvement and modernisation is long overdue. Although work to the cliff stabilisation at West Bay

Figure 1.3.18   *West Bay, Dorset*

is now complete, its long-term success must depend on a successful overall scheme for coastal protection and defence at West Bay, since damage to the Esplanade at the base would probably lead to a catastrophic landslip.

Cliff Stabilisation
The cliffs immediately to the west of the Esplanade at West Bay appear to be inherently unstable. They are made of Fuller's Earth Clay, which is traversed by a major **tear fault**, which has significantly weakened their structure (Figure 1.3.19). Flow of ground water within the fault zone may also have contributed to the overall instability. Landslipping seems to have occurred intermittently since 1969, and obviously poses a threat to the houses that are built on the top of the cliff (Figure 1.3.19).
In 1967 proposals were drawn up for a new sea wall and esplanade to the west of the existing structure in order to protect the cliffs from further erosion. Such work also involved a regrade of the cliff to an angle of 27°. In 1969 soon after the completion of the cliff regrade a slip occurred at the top of the cliff opposite the property known as Seahill (Figure 1.3.19). Although remedial work was carried out, with the building of drains, further slips occurred in the 1970s. Fresh work on the coastal defence at the base of the cliff was required from 1981–3 in order to repair damage done to the sea wall in 1977. Further remedial works were carried out in

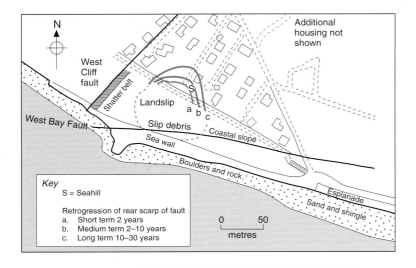

Figure 1.3.19 *Westcliff landslip, West Bay*

the late 1980s, but without seeming to solve the problem.

After a detailed engineering geologist's report, which suggested that if further works were not carried out, evacuation and demolition of at least one property would have to take place, a full programme of remedial works was carried out from 1995–6. Further drainage, and regrading of the slope was carried out, and its success will have to be judged by continuous monitoring over the next few years.

The New Coastal Defence Scheme
The new coastal defence scheme for West Bay is required to meet five technical objectives:
- reduce flooding of the properties behind the beach
- protect the beaches, cliffs and existing coastal defences from erosion
- protect against flooding around the harbour

- protect and reduce maintenance requirements both for the piers and the structure around the existing harbour
- maintain harbour access.

Three schemes were put forward, A, B and C, and these are shown in Figure 1.3.20.

Scheme A: The Remedial Scheme
Existing coastal defences would be strengthened. Existing rock bastions would be extended and supported by new bastions. Both East and West Beach would be nourished with material matching the existing beach material. Rock armour at West Pier would be extended.

Scheme B: Coastal Defences Incorporating a Rock Armour Breakwater
All of the works provided in /scheme A would be included in Scheme B, with the exception of the addition of rock armour to West Pier. This would instead be protected by extending a rock bastion to form the arm of an Outer Harbour.

Scheme C: Coastal Defence Scheme Incorporating Stone Harbour Breakwater
Scheme C is identical to scheme B, with the exception that the breakwater will be constructed either by concrete caissons or by a rigid superstructure with masonry wave wall on a rock breakwater. The breakwater would be provided with a walkway for public access, and eventually, berths for fishing vessels and leisure craft.

The Physical Environment
Existing coastal structures include the Harbour, the piers, the esplanade wall, rock armour, bastions and long sea outfall. All are liable to erosion and damage. Waves enter the Harbour through the gap between the two piers, which can give rise to flooding in the properties surrounding the Harbour. The beaches on either side of the Harbour entrance are both suffering retreat due to coastal erosion, and on East Beach this has led to the flooding of properties behind the beach.

Sediment is supplied to the area by local rivers (the Brit) and by erosion from the surrounding cliffs. Sediment transport is wave, rather than current generated. **Net drift** at East Beach was calculated in 1985 as being from east to west, but in 1991 local **wave climate** had changed and the drift was from west to east. Despite this

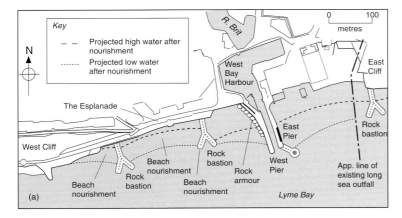

Figure 1.3.20 A *Coastal Defence scheme, West Bay*

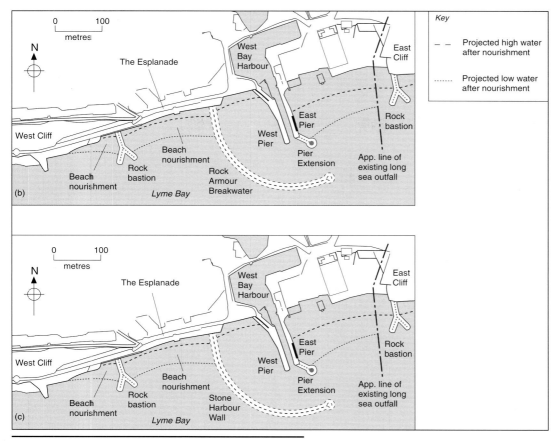

Figure 1.3.20 B and C   *Coastal Defence Scheme, West Bay*

there is still a significant build up of sediment on the eastern side of the piers, because net drifts are low and long-term impacts on the shoreline are slow to evolve.

Offshore, the seabed is characterised by muddy sands and gravels, and outcrops of limestone bedrock. Treated sewage is pumped offshore by Wessex Water via a long sea outfall running from East Beach.

The Natural Environment
West Bay is located within the Lyme Bay 'important area for marine wildlife'. There is a considerable diversity of community types, but species abundance and diversity are low because of environmental factors such as scouring by sand, mobile sediment and low levels of organic matter. Cliffs on either side of West Bay are included in the West Dorset SSSI (designated for its geological importance and the habitats supporting rare flora and fauna). Chesil Beach has SSSI status because of its value as a classic landform. West Cliff and East Cliff have SSSI status because of their fossil content, and structures.

The Human Environment
West Bay is an important tourist destination, with a range of the usual facilities. In-shore leisure activities include dinghy sailing, diving, water-skiing and board-sailing. There is a small fishing fleet at West Bay, operating mainly for in-shore fish and shellfish stocks. Facilities are somewhat limited, because of the narrow Harbour entrance, which, together with weather conditions, limits fishing to 150 days a year.

WHICH SCHEME SHOULD BE ADOPTED?

1 Review the major features of the physical and human environment of West Bay.
2 Compare the likely environmental impacts during construction and operation of the three schemes A, B and C.
3 Select the scheme that you consider would be most beneficial for West Bay.

ESSAY

What are the main physical problems that confront coastal managers in Dorset? Discuss the engineering solutions that are available, and, where appropriate, comment on their current level of success.

1.4 Managing the North East Coast of Ireland

THE GEOLOGY OF THE COAST OF NORTH EAST IRELAND

The geology of the coast is dominated by the Tertiary basaltic lavas that were erupted on a surface cut in Chalk. Although older rocks are encountered at several locations along the coast (Figure 1.4.1) it is the lavas and the Chalk that are responsible for most of the spectacular scenery encountered between the mouth of the River Foyle and Larne Harbour. The Chalk is remarkably different from the Chalk in southern England. In Northern Ireland it is exceptionally hard and impervious. The lava flows were erupted from fissures or volcanic vents, and were piled one on top of the other in gently dipping sheets.

Unlike Dorset, the coast of north east Ireland suffered the effects of glaciation. Most of the ice moved south from Scotland, although some local ice moved north from the Bann valley, the Glens of Antrim and the Sperrin mountains to reach the coast. Two main effects resulted from the ice. Large quantities of glacial debris were deposited around the coast, and the weight of the ice caused **isostatic subsidence** of 150 to 200 m. Sea level changes have fluctuated considerably over the last 20 000 years.

SHORELINE TYPES

A considerable variety of shorelines exists in Northern Ireland. A summary of the different types is shown in Figure 1.4.2.

Shore Type	Length (km)	Percentage of total coastline
Rock and earth cliffs	48.8	21.5
Rock platforms	35.9	15.9
Sandy beaches	29.7	13.1
Rock, boulder or gravel beaches	48.1	21.3
Artificial coasts (seawalls etc.)	52.1	22.9
Estuary/river mouth	12.1	5.3
Total	226.7	100.0

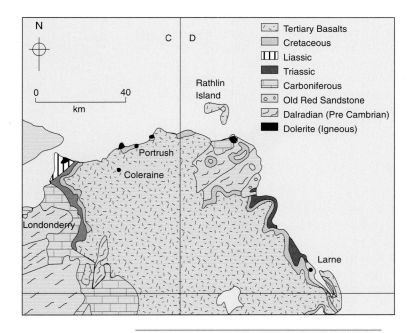

Figure 1.4.1 *Geology of coast of north east Ireland*

Figure 1.4.2 *Different shore types on the north east coast of Ireland*

SOME CASE STUDIES OF COASTAL MANAGEMENT IN NORTH EAST IRELAND

Coastal management in North eastern Ireland poses a similar range of problems to those in Dorset. Cliff instability is characteristic of much of the east coast of County Antrim, whilst loss of sediment from beaches is common to both the eastern and northern coasts. Once again balancing the needs of conservation and recreation is important in stretches that are heavily used by tourists, such as the length of coastline at the Giant's Causeway, and to the east. Figure 1.4.3 shows the current dynamic status of the coast of north east Ireland.

This section looks at;
- Portballintrae Harbour
- Managing Tourist Pressures: the Giant's Causeway
- Protecting the small beaches and settlements of eastern County Antrim.

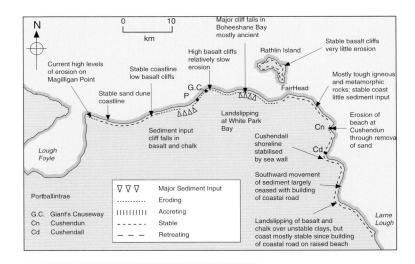

Figure 1.4.3 *Coast of north east Ireland current dynamic status*

PORTBALLINTRAE BAY

The evolution of Portballintrae Bay over the last 60 years might almost be said to be a case of 'mismanagement' rather than management! The bay was first formed about 5000 years ago at slightly higher sea levels than today. With sand influx from the adjacent shoreline and erosion from the cliffs at the head of the bay it developed a wide sandy beach, with a strip of gravel on the landward side. After the end of the nineteenth century, sand appears to have been lost at the rate of 1000 m³ every year.

The map (Figure 1.4.4) illustrates the essential features of the bay today. It seems that Leslie's Pier, near Seaport Lodge, built in 1895, is the basic cause of the problem. It upset the wave pattern in the bay, and caused bigger waves to break in the south west corner of the bay, this set up a strong southward drift to the centre of the bay. Sand began to move towards the centre, from whence it was moved off-shore into deep water. Groynes were built to stop the drift, but were soon destroyed by the waves. Larger waves began to attack the cliffs and erosion became a problem, with land being lost at a rate of 0.5 m per

annum by 1975. £250 000 has been spent over 25 years in various, unsuccessful attempts to stop this erosion. Today the beach has reached some sort of new equilibrium, as shown in the diagram, and it may be that this new grading may eventually halt the erosion.

MANAGING TOURIST PRESSURES: THE GIANT'S CAUSEWAY

The Giant's Causeway is a major tourist attraction in Northern Ireland. Records show that the first visitor's accounts and descriptions go back to the late

Boswell to Dr. Johnson: 'Is not the Causeway worth seeing?' Johnson's answer: 'Aye, worth seeing, but not worth going to see.'

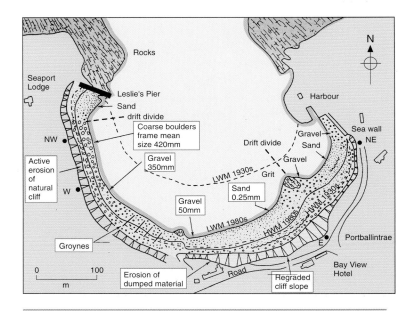

Figure 1.4.4 *Portballintrae Bay showing the sediment grading along the shore*

Figure 1.4.5 *The Causeway Centre, Giant's Causeway, Northern Ireland*

seventeenth century. For over 200 years visitors to the Causeway were entertained and instructed by groups of unofficial casual guides. After World War II it became increasingly obvious that management of the site would have to be rationalised if the Causeway and its surrounding area were not to suffer lasting damage.

Since the 1960s, management of the Causeway, and the adjacent cliffs, has had the benefit of a series of designations, all of which aid conservation in such a 'honey-pot' site.

1 1961–4: acquisition of the site by the National Trust, together with much of the adjacent coastline

2 1986: it was accepted by the World Heritage Convention on to its list of sites. The Causeway meets two of the criteria for an outstanding natural property
   ● it is a prime example of the earth's evolutionary history during the Tertiary Era
   ● it contains rare and valued natural phenomena

3 1987: the Giant's Causeway, and 71 hectares of the adjacent coastline were designated a National Nature Reserve.

4 1989: the coast of Northern Ireland between Portrush and Ballycastle was designated an Area of Outstanding Natural Beauty.

A clear focus for the interpretation of the Giant's Causeway has now been established in the Causeway Centre opened in 1986 (Figure 1.4.5). Other important management initiatives are the provision of transport from the Centre to the Causeway itself, and the creation and upkeep of a series of coastal paths, with appropriate interpretive facilities.

STUDENT **1.4** ACTIVITY

Suggest how the diagram of the coast at the Giant's Causeway (Figure 1.4.6) might be used as an interpretive board on the coastal path.

PROTECTING THE SMALL BEACHES AND SETTLEMENTS OF EASTERN ANTRIM

At the head of the small bays on the east coast of Antrim there is a series of small beaches that appear to be losing sediment, largely through some form of human interference. In several cases this appears to be the result of sand and gravel being taken from the beaches by farmers. In the past beaches could cope with small amounts of sand being removed, but in recent years quantities have increased with improved methods of transport and equipment. It is becoming clear that unless these beaches are managed properly, they are likely to disappear altogether. Figure 1.4.7 shows the beach at

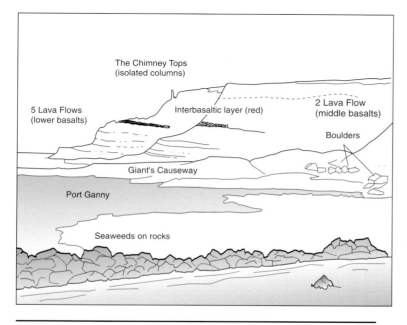

The Chimney Tops
(isolated columns)

5 Lava Flows
(lower basalts)

Interbasaltic layer (red)

2 Lava Flow
(middle basalts)

Boulders

Giant's Causeway

Port Ganny

Seaweeds on rocks

Figure 1.4.6 *Physical features in Causeway bays and headlands looking east*

Cushendun, the most northerly of the little settlements at the head of the bays. The beach appears to have been derived mainly from river material brought down by the small stream that flows into the southern end of the bay. The diagram illustrates the likely circulation of sand within the bay. Attempts to dredge or channelise the river in the past may have interfered with the circulation cycle, steadily cutting the river mouth off from the beach.

Apart from this disruption in the cycle, the removal of increasing amounts of sand from the beach has further complicated the situation. The removal of sand has been seen as a right possessed by farmers, embodied in Common Law, and handed down from one generation to another. About 300 loads of sand are removed from the beach each year, and this could mean that since 1960 45 000 m³ have been lost – equivalent to the amount of mobile sand on the beach today. Under these conditions the beach takes sand from the backshore cliffs and retreat occurs. Before 1960 retreat was relatively slow in the bay – only 0.1m to 0.2m per year, but in more recent times erosion has accelerated, increasing to as much as 0.8 m a year at the south end of the beach. Management of the beach to stop the loss of sand is now clearly needed, and all the involved authorities – the Crown Estate Commissioners, the National Trust (who lease and manage the beach) and the Department of the Environment are currently examining management proposals. Supplying non-beach sand to local farmers is seen as a possible solution.

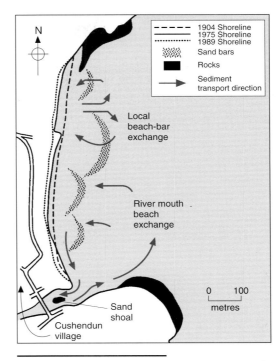

Figure 1.4.7   *Cushendun Bay*

SUMMARY ESSAY

To what extent has human interference with coastal processes been responsible for erosion problems along the coast?

FIELDWORK OPPORTUNITIES & PROJECT SUGGESTIONS

1 Select a stretch of coastline which has serious physical management problems. These could be erosion, landslipping or flooding (or a combination of these problems). Your fieldwork project could
   a assess the nature of the processes that are causing the problem
   b identify and map the management initiatives and techniques that are being used to deal with the problem
   c Assess the effectiveness of the measures taken.

2 Select a coastal area, like Lulworth Cove, or the Giant's Causeway which is under severe tourist pressure. A project could be devised to
   a map all of the physical features that attract tourists to the piece of coastline
   b locate all of the areas of tourist/environment stress (where the environment is at risk from tourists), and attempt to classify them e.g. restricted access to beach, footpath erosion
   c assess the effectiveness of the management techniques used to reduce the level of stress.

2

The Management of Rivers

The fact that rivers are such a symbol of endurance and changeless change is what makes their management a touchstone for the whole issue of our relationship with the natural world

Jeremy Purselove: Taming the Flood.

KEY IDEAS

- River management depends on a clear understanding of the main processes at work in the basin catchment
- An **holistic** approach to river management is essential
- Strategies need to incorporate natural river processes into any scheme designed for the efficient, but sensitive management of the basin
- Conflicts between different users need to be reconciled in any successful programme of river management
- River management must be sensitive to the ecological needs of the system

2.1 Introduction

The overall picture presented by the management of rivers and estuaries in Britain and Ireland is, like coastlines, complex, and somewhat confusing. In England and Wales the work of the National Rivers Authority (NRA) (now incorporated into the Environment Agency), with a wide range of responsibilities, involves a co-ordinated approach to river management. In Scotland, management of rivers is shared between a number of authorities, both local and national. The situation in the Republic of Ireland is broadly similar, but in Northern Ireland responsibility for river management is largely in the hands of the Department of Agriculture and the Environment.

The NRA was created through the Water Act of 1989. It has proved to have a relatively short life, since, in April 1996, it was subsumed within the much larger Environment Agency. Put simply, in its own words, the National Rivers Authority's role is 'to protect and improve the water environment of Britain'. With its responsibility for coastal waters and groundwater, its remit goes far beyond river management. However, in its twelve stated aims, river management is mentioned or implied in nine. The Environment Agency has seven areas of responsibility for rivers, as shown in Figure 2.1.1

THE RIVER BASIN

Management of rivers increasingly tends to take a holistic view, regarding integrated action for the whole of the drainage basin as being the essential way in which rivers may be successfully controlled. In using the basin approach to management, all of the varying strands of strategy, indicated in Figure 2.1.1 can be combined.

Figure 2.1.2 shows a threefold division of a drainage basin into three zones: source, or sediment production zone, the 'transfer' zone, and the depositional zone. Figure 2.1.3 shows the change in channel properties through these three zones.

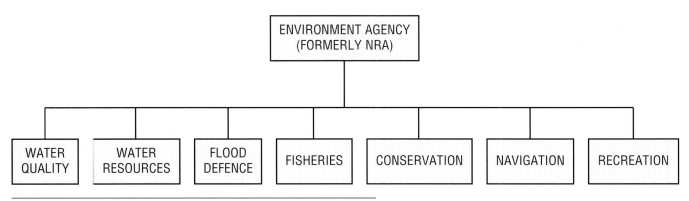

Figure 2.1.1   *River management responsibilities of the Environmental Agency*

STUDENT   2.1   ACTIVITY

1 Critically review the value of Figure 2.1.3
2 Explain carefully the relationships shown in Figure 2.1.3
3 Explain why a thorough understanding of these relationships is essential for successful river management.

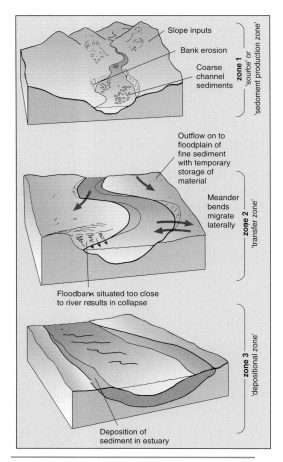

Figure 2.1.2   *Variation in channel morphology and processes through a drainage basin*

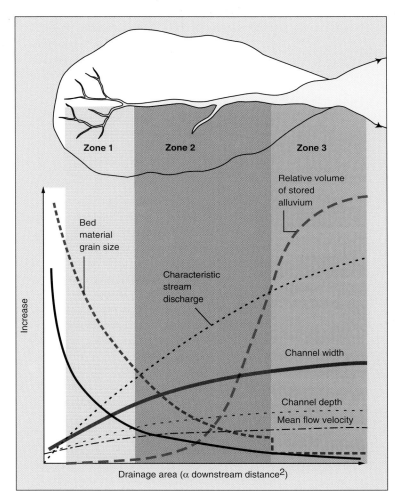

Figure 2.1.3   *Changes in channel properties through a drainage basin*

Traditional engineering responses to river management (Figure 2.1.5) have not always been sympathetic to the **geomorphology** and ecology of the river channel and its banks. Rivers are now seen as dynamic systems, in which flooding is a natural process. It is now recognised that

Engineering objective	Impact
Reduce period of flooding on adjacent land and provide better channel to enable drainage	(a) Degraded channel morphology with loss of pool/riffle habitats. Channel destabilised upstream and downstream. (b) Decreased potential for backwater creation/rejuvenation. (c) Bankside plants lost because of steep and rapid transition from permanently wet channel to dry banks. (d) Water table lowered in floodplain habitats adjacent to river leading to loss/reduction of wetland species.

Figure 2.1.4 *Potential ecological and geomorphological impacts of river deepening*

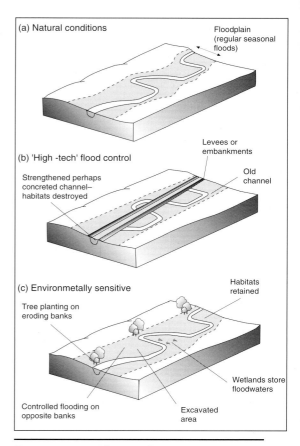

Figure 2.1.5a *Changing ideas on river management*

traditional engineering alteration to channel width will cause changes in water velocity and sediment transport. Channel deepening, too, is often seen as a way of reducing floor risk within a stretch of river. Figure 2.1.4 shows some of the ecologically damaging effects of deepening. The straightening of rivers has often resulted in a very variable discharge. Healthy river ecology depends essentially on a diversity of habitat which requires the retention of natural features of the river, including pools and riffles, overhanging river banks, and a range of sediment sizes in the stream. Three essential strategies are seen as key elements in the management of rivers (Figure 2.1.5a):

- new river channels should be based on the geomorphology and ecology of natural and stable river channels. Sensitively designed meandering channels are to be preferred to artificially straightened channels;
- low impact techniques should be used in future river management. For instance, the planting of trees and shrubs along banks liable to erode can be very effective;
- river corridors, which allow the river to function as naturally as possible, are seen

as the preferred management option. Corridors several times the width of the river channel will allow the stream to meander, and permit controlled flooding in certain sections.

Figure 2.1.5(b) shows how geomorphological solutions are available as a response to change within the river basin. These are basically alternatives to traditional engineering solutions.

STUDENT **2.2** ACTIVITY

1 Select any small stream or river in your own home area, and map a section to show what traditional engineering methods have been used in response to problems that need management.
2 Suggest alternative geomorphological responses that could be used to promote a more harmonious relationship with river and bank ecology.

Activity within catchment	Impact on channel	Traditional engineering response	Geomorphological response
Urbanisation	Lowered base flows and increased flood flows from decreased infiltration and lowered groundwater levels. Increased sediment load initially. Increased run-off from paved areas – increased discharge – widening through erosion (especially in non-cohesive glacial sands and gravels). Overwide channel encourages deposition of urban silts.	Enlarge channel. Line channel with concrete or similar armour layer. Sheet piling or concrete on both banks.	Source control of run-off using permeable pavements, storage ponds etc. Establish a two or multi-stage channel to accommodate increased discharges while maintaining a low flow width. Sequence of pools and riffles placed in channel to enhance habitat. Wetland flora can be planted on higher berm. Erosion control using geotextiles (if necessary).
Realignment/ Diversion to accommodate development	Loss of pool/riffle sequence. An increase in slope – higher velocity – erosion of bed and banks – instability – wider, more uniform channel. Decrease in ecosystem habitat.	Straighten channel – shortest point A–B uniform, trapezoidal morphology.	Realignment should copy original plan geometry. Minimise slope increase and reinstate bends. Pools and riffles reinstated at regular intervals. Reinstate substrate.
Bank Protection	Loss of natural bank profiles. Loss of marginal habitats.	Toe boards, sheet piling, sandbags.	Planting of reeds, geotextile stretch fencing, gabions or rip-rap, willow piling. These should be placed on the outside of a bend where flowlines converge.

Figure 2.1.5b   *Traditional engineering and geomorphological responses to changes in drainage basin*

CATCHMENT MANAGEMENT PLANNING

As a result of the need for integrated planning for drainage basins, the NRA introduced the concept of Catchment Management Planning. This treats a particular river, together with its tributaries and groundwater resources as a whole unit. Decisions taken on one aspect of river management are related to other facets. For example, flood control and alleviation may well affect fisheries or conservation, and any strategy must consider all three. Catchment Management Plans operate in two stages. An initial Consultation Report recognises the range of management issues in any one catchment, and outlines options for action. It is freely available to all interested organisations and the general public, who are invited to comment on its content. The Final Report follows after the consultation period, and includes an

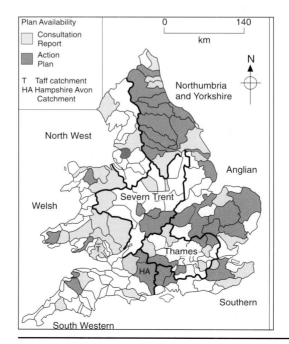

Figure 2.1.6   *Environment Agency catchment management plans*

Action Plan, which indicates how the Environment Agency intends to address the management issues identified. The Environment Agency states the aims of the Catchment Management Plans as:
- translate EA functional policy into integrated catchment policy
- prioritise needs for the water environment
- sponsor integrated action and sustainable development
- facilitate understanding and ownership of multi-functional working
- establish a means of fair and equitable consideration of all water users.

Within England and Wales there are 163 catchments, and plans should be in place for all of these by the end of 1998. The map in Figure 2.1.6 shows the current level of preparation of Consultation Reports and Action Plans for the catchments in England and Wales.

2.2 Case Studies of the Management of Rivers

Two case studies of Catchment Management Plans are selected for examination. The Hampshire Avon has a catchment area of some 1700 km², whilst the River Taff in South Wales is somewhat smaller at 526 km². The catchments of the two rivers show some sharp contrasts when examined. The Hampshire Avon, although it rises on the Upper Greensand in the Vale of Pewsey in North Wiltshire is basically a Chalk river (Figure 2.2.1), in so far as nearly half of its course runs across the Chalk, and it receives most of its tributaries from the Chalk. The Taff, by comparison, rises as a mountain stream (Figure 2.2.2), high on the sides of the Brecon Beacons, and flows southwards through the coalfield areas of South Wales

Figure 2.2.2 *The Taf Fawr, near Storey Arms, Brecon Beacons*

Figure 2.2.1 *The Hampshire Avon, near Bodenham, Wiltshire*

to enter the Bristol Channel at Cardiff. The Hampshire Avon flows mainly through a rural area, with Christchurch, near its mouth, being the largest **riparian** town. In contrast, the Taff flows from the Brecon Beacons National Park, to the north of Merthyr Tydfil, through the industrial areas of South Wales.

The work of the Environment Agency is much involved with the provision of flood defences. The River Stour in Dorset has a long history of flooding, and small towns such as Blandford and Christchurch have been particularly affected. In recent years the NRA has been responsible for comprehensive flood control and alleviation work at a number of vulnerable locations along the Stour.

2.3 The Basin of the Hampshire Avon

Figure 2.3.1 shows the catchment of the Hampshire Avon. The map shows that the main groundwater supplies to the river come from the Chalk aquifer, which yields mainly alkaline water. Important supplementary supplies come from the Upper Greensand, although these waters are more acidic. Since the Avon is spring-fed from Chalk, it tends to have a fairly stable discharge, with an average flow of 19.7 cumecs. Generally the catchment responds slowly to rainfall, and flood events are rare (1960 and 1990 being exceptional circumstances).

STUDENT 2.3 ACTIVITY

Study the yearly hydrograph of the Hampshire Avon at Knapp Mill, near its mouth at Christchurch (Figure 2.3.2).

1 Describe and explain the general form of the long term average hydrograph.
2 Attempt to account for the variations that occurred in 1992.

The Avon flows through an area which is predominately rural in character, with fairly intensive farming dominant. However, within the catchment there is a relatively high proportion of semi-natural habitats compared to others in southern England. The map, (Figure 2.3.3) shows the land use in the river corridor of the Hampshire Avon. It shows a dominance of pasture land, together with the semi-natural habitats of woodland, scrub and marsh. However, since these uses are essentially riparian, i.e. within 25 m of the river bank, the map does not give a true picture of the overall land use in the catchment. Away from the floodplain of the river, arable land does become very important, particularly so on the Chalklands of Wiltshire.

Most of the upper catchment, upstream from Fordingbridge, is used for arable, sheep and dairy production (Figure 2.3.4). This has important implications for water abstraction, water quality and land drainage. Nitrate leaching and increased sediment input into the catchment are

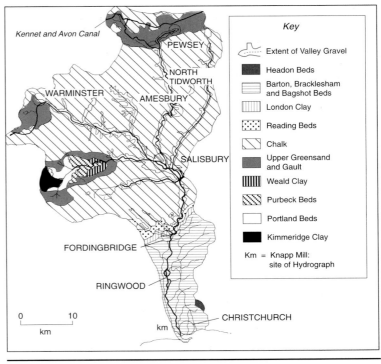

Figure 2.3.1 *Hampshire Avon catchment geology with principal aquifers (chalk and Upper Greensand)*

obvious results of the intensification of farming. Concern has been expressed that there has been a degree of habitat loss in the upper catchment, both within the river channel and on adjacent land. By comparison, habitat loss has been much less in the lower part of the catchment, where considerable parts of the valley are protected and managed along more traditional lines, although few of the original water meadows remain except near Salisbury.

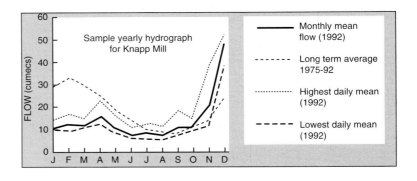

Figure 2.3.2 *Sample yearly hydrograph for Knapp Mill*

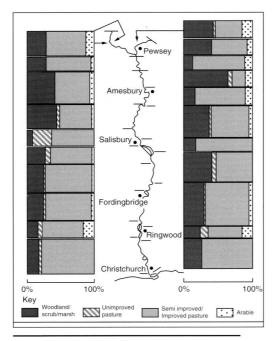

Figure 2.3.3 *Hampshire Avon catchment river corridor, land use*

Key: Woodland/scrub/marsh — Unimproved pasture — Semi improved/Improved pasture — Arable

Figure 2.3.4 *The upper Avon valley near Durrington, Wiltshire*

THE HAMPSHIRE AVON: THE MAIN MANAGEMENT ISSUES

Within the catchment of the Hampshire Avon, four main management issues may be identified;

- Water Quality
- Water Resources
- Fisheries
- Multi-functional Conflicts.

Water Quality

Water quality in rivers is now expressed in ecosystem classes. Those for the Hampshire Avon are shown in Figure 2.3.5.

It would appear from Figure 2.3.5 that the Hampshire Avon at present possesses high standards of water quality, both in the main trunk river and its tributaries. However there are still matters of concern that the Environment Agency wishes to address. The principal Sewage Treatment Works in the catchment are shown on Figure 2.3.6. Water quality downstream from the works at Warminster, Salisbury, and Ringwood does need improvement. Farm discharges in the Nadder catchment have adversely affected river quality, as reflected in the lower River Ecosystem Classification (Figure 2.3.5).

Pesticides (mainly triazine, which has been used in agriculture and general weed control) have been detected at the Water

Abstraction plants at Matchams and Knapp Mill. Their main source would appear to be in the Nadder catchment. Triazine is used as a pesticide for maize, which is grown widely as a fodder in the catchment, and this is likely to continue. Watercress farms are responsible for some pollution input into the Avon.

Eutrophication does not appear to be a major problem in the Avon, although there are some signs that it is moving towards the higher levels that are seen in the neighbouring River Stour, in Dorset.

Water Resources

Surface and groundwater from the Avon Catchment are abstracted for a variety of

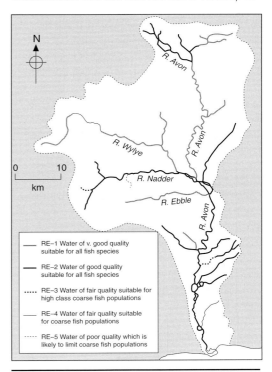

Figure 2.3.5 *Hampshire Avon, water quality, river ecosystem quality*

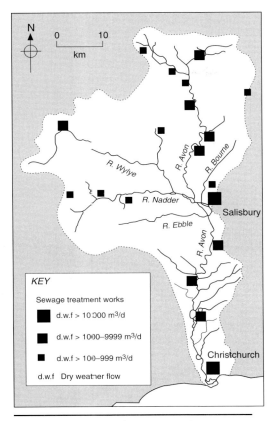

Figure 2.3.6 *Hampshire Avon catchment sewage treatment works*

purposes, including public water supply, spray irrigation, aquaculture (fish farming), gravel washing and industrial uses (see figure 2.3.7). Much of it is returned at some point in the catchment. Groundwater abstraction occurs principally in the upper catchment, from the Greensand and Chalk aquifers north of Salisbury. In recent years this has given cause for concern, simply because of low flows recorded in the Upper Avon, the Wylye, the Nadder, the Bourne and the Ebble.

In the Lower Avon, public water supplies are abstracted from the river at three locations – Blashford Lakes, Matchams and Knapp Mill (see Figure 2.3.7). These three locations are responsible for nearly 18 per cent of the total water extraction from the Avon, and demand is set to increase considerably as a result of population growth in south east Dorset. Non-consumptive abstraction, such as in fish farms, in fact returns water to the river usually very close to the point of abstraction. Fish farms are responsible for nearly 80 per cent of the total surface

water abstracted in the catchment. Figure 2.3.7 shows the total licensed surface abstraction from the Avon.

Fisheries

The Hampshire Avon is one of the major salmon fishing rivers in southern England, and sea trout fishing is also important in the lower Avon around Christchurch. Concern has been expressed over the decline in salmon numbers, which can be traced back over 40 years, but seems to have accelerated recently. A variety of reasons have been put forward for this decline, including patterns of exploitation by anglers, changes in sea temperatures, and reduced river flows in the early 1990s. In the Upper Avon brown trout fisheries are of national importance, with the physical and chemical properties of the water providing an ideal habitat for the fish and for angling. Fishing quality appears to be declining in the Upper Avon also, and this may be attributed to loss of weed, loss of water clarity, and siltation.

Multi-functional Conflicts

Conflict between abstraction of water on the Lower Avon, and flow requirements for the upstream migration of salmon: with the likely increase in water abstraction from the

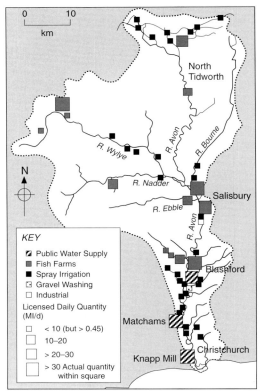

Figure 2.3.7 *Hampshire Avon catchment surface water abstractions*

Lower Avon, consequent upon population growth in south east Dorset, concern exists that lower flows may disrupt the upstream migration of salmon. A 1986 study showed that there was a clear relationship between salmon movements into and up the river and the level of flow.

Weed-cutting in the Avon: can have a number of effects on the functioning of the river system. In many rivers flow levels tend to fall when weed growth commences, but in the Avon this is not the case. Without weed cutting overbank flow is likely to occur, with an obvious effect on haymaking in the water meadows in riparian holdings. Weed-cutting can significantly reduce water levels in the river, and this, in its turn can seriously affect wader bird populations. Weed-cutting is, however, beneficial to both game fishermen, and coarse fishermen.

Development control within the catchment of the Avon: is clearly of importance to the Environment Agency. However it has somewhat limited control over current land use changes and allocation, within the catchment. All three County Structure Plans (Wiltshire, Hampshire and Dorset) include policy statements that are relevant to the management of the Avon catchment. New roads, such as the planned Salisbury by-pass (see Chapter 10, page 175), will have important implications for both water quality and water quantity, and conservation and landscape of the river corridors that they cross. Developments within the catchment, and, in particular within the flood plain, can present a flood risk, either in the immediate area, or elsewhere on the flood plain.

Conservation and recreation: have the potential to create conflict within the Avon catchment. The Avon valley below Netheravon is an Environmentally Sensitive Area, which offers incentives to farmers to practise traditional methods of farming, maintaining a high summer water table, and increasing the frequency of shallow flooding of riparian meadows in winter and early summer. There are over 60 SSSIs, and it also qualifies as a Special Protection Area (SPA) and its wetlands have **Ramsar site** status.

Although water-based recreation is relatively limited on the Avon, Christchurch Harbour is used for a wide range of water sports, and lakes within the catchment are used for similar purposes. The Avon Valley is seen as having a high amenity value, and the potential for conflict between organised and casual recreation, and conservation interests has to be sensitively managed in the future.

STUDENT ACTIVITY

1 You have been commissioned as an independent adviser to the Environmental Agency, and have been asked to comment on current issues and conflicts in the Avon Catchment area.
   a Summarise the main issues that currently confront the managers of the Avon system.
   b Prioritise the order in which these issues should be addressed.
   c Draw up a conflict matrix showing how different uses of the system are likely to cause difficulties.
   d Suggest how these conflicts might be resolved.

2.4 The Catchment of the River Taff, South Wales

The map (Figure 2.4.1) shows the nature and extent of the Taff catchment in South Wales. It includes not only the Taff, but also its important tributaries the Cynon, Rhondda, Clydach and the Taf Bargoed.

Although the Taff rises as an upland stream (see Figure 2.2.2), much of its catchment embraces the basin of the South Wales coalfield, and its former industries of coal, iron and steel. Such industries had an

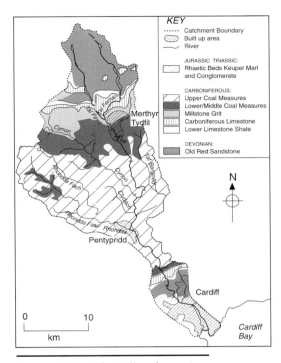

Figure 2.4.1   The Taff catchment Area

important effect on water quality and discharge in the Taff and its tributaries, but with their almost total demise in the twentieth century, water quality has improved, and the ecological status of the river has been enriched. Compared to the Hampshire Avon the Taff catchment receives very heavy rainfall, which increases upstream from 950 mm at Cardiff to 2400 mm in the Brecon Beacons. The high rainfall, combined with the upland catchment and steep gradients of the river channels, contributes to high and rapid discharges in the catchments. Unlike the aquifers in the Hampshire Avon catchment the rocks and soils in the Taff catchment have only modest storage capacity, and thus in periods of low rainfall flows can recede quite rapidly. Average flow of the Taff is slightly less than that of the Hampshire Avon at 18.63 cumecs. Unlike the Hampshire Avon, the Taff has sizeable reservoir systems in its upper catchment, on the Taf Fawr, and on the Taf Fechan. These reservoirs have a significant effect on the flow in the Taff, particularly in winter, when flood waters are stored, and flow levels in the Taff are reduced. Groundwater levels in the Taff catchment are highly influenced by the pattern of old mine workings, through which a complex pattern of flow has developed.

Land use varies within the Taff catchment. In the upper catchment much of the Brecon Beacons is open moorland, although there are considerable areas of forestry plantations. Unrestricted run-off on the moorland slopes has led to intensive gullying in the past. There are almost 100 km^2 of forestry plantations within the catchment (20 per cent of the total catchment area). Adverse effects from forest can include:
- increased sediment load and run-off rate in the initial stages that can increase flood defence maintenance, and can destroy conservation features;
- reduced water yield as trees intercept more rainfall;
- enhanced acidification as trees capture more acid pollutants from the air;
- pollution from fertilisers and pesticides applied to the crop.

Farming in the upper catchment is principally sheep farming, with the occasional dairy beef or pig farm. It appears to have little impact on water quality. In the lower parts of the catchment, the Taff and its tributaries are heavily urbanised, particularly on the valley floors, and this has led to increased run-off patterns. Thus flooding has been a major problem on many of the riparian settlements, although flood defences are now in place in many of the affected locations.

Numerous derelict collieries, steel works and spoil heaps remain as a legacy of the industrial past of the area. Land reclamation has occurred, or is in progress, at many of these sites, and the disturbance may cause serious pollution of rivers due to the release of suspended solids and the mobilisation of **leachable components**. Ferruginous minewater from old abandoned mines is another problem, resulting in the discoloration of water and the deposition of iron hydroxide on the river bed. The last, recently reprieved deep mine, Tower Colliery at Hirwaun, in the north of the coalfield, still discharges washery waste, minewater and site drainage into the headwaters of the River Cynon. Some opencast mining remains: site drainage from this is often acidic, and is neutralised before being discharged into the river system.

Some quarrying remains within the Taff catchment, principally the limestone quarries on the northern and southern rims of the coalfield. Waste streams from these quarries tend to be alkaline, and suspended solids and oil leakage are other difficulties that are associated with these workings. Most industrial process effluent from the industrial estates that have replaced the heavy industries is now discharged from the works to the foul sewer, or direct to coastal waters.

MANAGEMENT ISSUES IN THE TAFF CATCHMENT

For the purposes of comparison, the main management issues in the Taff catchment may be listed similarly to those in the Hampshire Avon:

- Water Quality
- Water Resources
- Fisheries
- Flooding

Water Quality

The map, (Figure 2.4.2), shows the state of water quality in the catchment of the Taff and its tributaries.

STUDENT **2.5** ACTIVITY

1 Compare Figure 2.4.2 with the equivalent map for the Hampshire Avon.

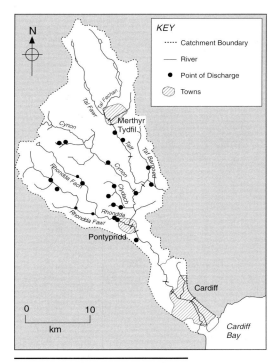

Figure 2.4.3 *Minewater discharges*

a What are the principal differences in water quality between the two catchments?

b How might you seek to explain these differences? Figures 2.4.3 and 2.4.4 provide important information on which you can make your judgements.

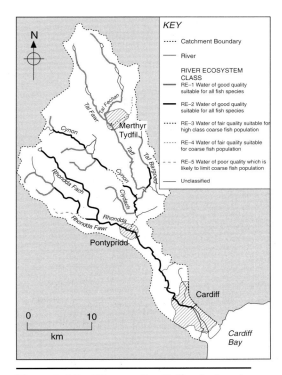

Figure 2.4.2 *Water quality, Taff and tributaries*

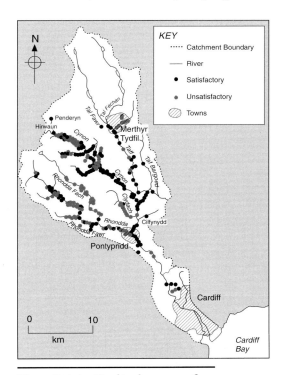

Figure 2.4.4 *Combined sewer overflows*

Other water quality issues identified by the Environmental Agency in South Wales are:

- the impact of road developments at Pontypridd and north of Merthyr Tydfil (the extension of the A470 here will pass through old mine workings);
- terrestrial and sewage litter in the Taff and its tributaries;
- the impact of land reclamation schemes on the water environment;
- lack of adequate legislation to control diffuse/intermittent pollution from industry;
- the impact of unsatisfactory treatment at the Cynon and Cilfynydd sewage treatment works;
- housing developments at Hirwaun and Penderyn that will further exacerbate difficulties in an antiquated and overloaded sewerage system.

2 Discuss how these issues may have contributed to river ecosystem class failure in the Taff catchment (failure to reach a target level set by the Environment Agency).

Water Resources

Interestingly, the level of water use in the catchment has seen a decline from the peak in 1970 (650 Ml/d) to 300 Ml/d in 1975, although it has since risen again to 335 Ml/d in the early 1990s. The principal reason for this drop in consumption has been the decline of manufacturing industry. The main use of water abstracted from the Taff and its tributaries is for public water supply (65 per cent of the total water abstraction). Ninety seven per cent of this comes from surface water sources (see Figure 2.4.5). Seventy-five per cent of this water comes from the reservoirs in the Taf Fechan and the Taf Fawr. Water from a number of sources in south east Wales can be redirected to locations where there is a shortfall. Abstractions for agriculture, fish farming and industry are relatively small, and the latter is increasingly taking its supplies from the mains.

Fisheries

Before coal mining and iron and steel production commenced in the late eighteenth century, the Taff was a prolific salmon river. Inevitably with the levels of pollution in the nineteenth and twentieth centuries its value as a salmon fishery

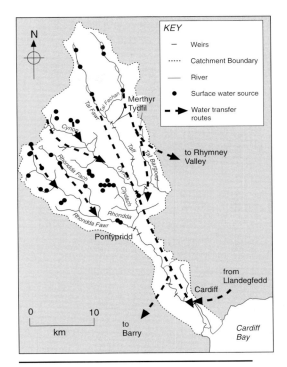

Figure 2.4.5 *Water abstraction, Taff catchment*

declined markedly, but, with improvements in water quality in the latter half of the twentieth century, salmon began to reappear in the 1980s. Salmon are now, with sea trout, found naturally in the Taff downstream from Pontypridd. Brown trout are particularly important in the upper catchment above Pontypridd.

Flooding

Flooding in the Taff catchment has been a problem since the urbanisation of the valley floors occurred in the late nineteenth and early twentieth centuries (Figure 2.4.6). The main causes of this flooding are:

- the heavy rainfall in the catchment, particularly in the upper catchment;

Figure 2.4.6 *Urbanisation on valley floor, Rhondda Valley, South Wales*

- the steep gradients and constricted channels of many of the streams within the catchment;
- gullying and land use changes in the upper catchment in the Brecon Beacons;
- urbanisation on the valley floors, with the creation of impermeable surfaces that has increased surface runoff rates;
- the clogging of parts of the course of rivers and streams with waste and rubbish.

The map (Figure 2.4.7) shows the main flood alleviation schemes that have been built in the Taff catchment area. In the Rhondda Fawr valley much residential housing for miners was built on low-lying, valley-bottom sites, often just upstream from constrictions in the valley. Hence floodwaters impounded by the constrictions would readily flood exposed areas upstream. In 1979, severe flooding occurred in a number of locations in the Rhondda Fawr, including Blaencwm, Gelli, Hopkinstown, and Pontypridd (where the Rhondda Fawr enters the Taff itself). Low lying property was particularly vulnerable.

In Pontypridd itself (Figure 2.4.8) Sion Street suffered flooding on almost a two yearly frequency in the 1980s and early 1990s and the town centre suffered a similar fate. Schemes of flood alleviation

Figure 2.4.8   *The River Taff at Pontypridd with Sion Street on the right*

were competed for the two areas in 1993 and 1994. Despite the number of flood alleviation schemes in existence, some areas remain at risk.

The Impact of the Cardiff Bay Barrage on the Taff System

The Cardiff Bay Barrage (see Figure 2.4.9) will cross the mouth of Cardiff Bay from Penarth Head to Alexandra Dock, impounding both the Taff and the Ely, and creating a freshwater lake. Such a development will have far reaching effects on the Taff Catchment. Water quality standards within the impounded body will require a range of management initiatives to maintain the required levels. Direct discharges of crude sewage will have to be diverted, and major combined sewer overflows will require substantial modifications. Dissolved oxygen levels will have to be enhanced, and **nutrient stripping** may be necessary upstream to prevent **algal blooms** in the impounded water within the Barrage. Other likely impacts are:

- flood alleviation will be enhanced in the low-lying areas of Cardiff;

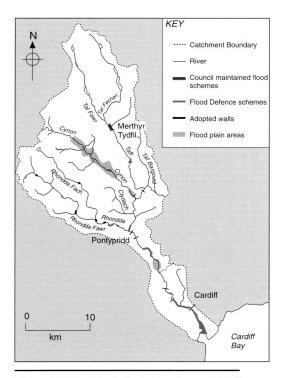

Figure 2.4.7   *Flood alleviation schemes in Taff catchment*

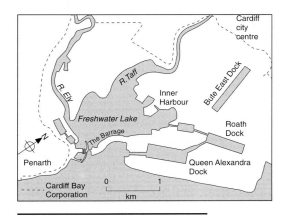

Figure 2.4.9   *The Cardiff Bay Barrage*

- the barrage will increase siltation in the Taff Estuary, and the removal of this material will be the responsibility of the Cardiff Bay Development Corporation (CBDC);
- the Site of Special Scientific Interest within the Barrage will be inundated, but it is thought that the water will develop special ecology;
- a special fish pass will be included within the Barrage. Concern has been expressed that the passage of certain groups of fish may be hindered. The efficacy of this fish pass will be constantly monitored by CBDC;
- regeneration of natural salmon and sea trout populations may be delayed;
- the freshwater lake will provide a very important coarse fishery;
- non water-contact recreational activity will develop within the impounded lake, and conflicts of interest may develop. Water-contact recreational activities are unlikely to develop because of the inadequate levels of water quality.

STUDENT   **2.6**   ACTIVITY

1 Using all the information and resources in sections 2.3 and 2.4, write a full comparative report on the principal management issues in the Hampshire Avon and Taff catchments.
2 Outline the range of common issues that managers of the two catchments might wish to discuss at an inter-catchment conference.

2.5 Flood Alleviation on the River Stour

The River Stour (Figure 2.5.1) in Dorset rises at Stourhead near the Wiltshire border, and then flows south eastwards for some 90 km. Its catchment is some 1.291 km^2 and embraces a wide range of geology (Figure 2.5.2). In its upper reaches the Stour and its tributaries flow across Blackmore Vale, floored with Kimmeridge Clay, Corallian Limestone and Oxford Clay. The Stour then flows across the outcrop of the Chalk, where it is fed by two important tributaries, the Allen and the Winterbourne. In this section it is also fed directly by a number of powerful springs from within the Chalk. In the

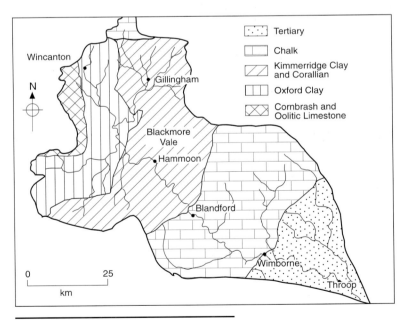

Figure 2.5.2   *Geology of the Stour catchment*

lower part of its course it flows across the sands, clays and gravels of the **Tertiary** Beds before entering the sea at Mudeford. The total fall from source to mouth is some 150 m, but gradients in the lower reaches are particularly shallow, being of the order of .0006. Flow does tend to vary a great

Figure 2.5.1   *The River Stour at Iford, Dorset*

Blandford Forum is the best, most complete small Georgian town in England, and its setting is superb, with water meadows all along the river, and on the south towards Bryanston a steep wooded slope known as The Cliff. Seen from the hills the town looks like a tiny Italian city.

Jo Draper: Dorset, the Complete Guide

deal on the Stour, despite the fact that it does cross the Chalk outcrop for a section of its course. Maximum flow recorded at Blackwater Bridge at Christchurch is 212 cumecs, dropping in the drought of 1976 to 1.12 cumecs, with an average summer flow of 2 cumecs.

The Stour has had a long history of flooding, both in the upper reaches in the neighbourhood of Gillingham, and also in the vulnerable areas of riparian settlements, such as Blandford, Wimborne, and Christchurch. Serious floods were recorded in the upper catchment in 1894 and in 1917, and in the lower catchment, Christchurch has suffered serious flooding on a number of occasions, in 1954, 1960, 1966, twice in 1979 and in 1990. The main causes of flooding appear to be:

- the dense drainage network in Blackmore Vale
- high surface run-off on the impermeable clays of Blackmore Vale. Most of the channel cross-sections in this lowland are inadequate to contain the twice per year flood event. Although they were improved in the 1930s and 1940s most of them are now in a poor state of repair
- rapid response to heavy rainfall in the upper reaches of the catchment
- exceptional high tides within Christchurch Harbour
- obstructions of flow by railway and road crossings and weirs

- urban development associated with the continued growth of the Poole-Bournemouth-Christchurch conurbation (500 000 people by the year 2000) has increased runoff in the catchment of many tributaries of the Stour, such as the Moors River
- the use of the floodplain for refuse dumping in the early twentieth century. This has raised levels and thus prevented the floodplain from acting as a natural temporary retention basin for flood water.

Flooding has been exacerbated by the lack of planning controls on development prior to the 1970s. This has meant that significant residential and commercial development has occurred on the flood plain, particularly in the lower catchment at Christchurch. Not only has this meant that property is vulnerable, but it has enlarged the area of impermeable surface, that leads to increased runoff.

One flood alleviation scheme, at Blandford Forum, is studied in detail below.

THE BLANDFORD FORUM FOOD DEFENCE SCHEME

Blandford Forum lies some 25 km to the north west of Bournemouth. It is built on the left bank of the River Stour, and has had a recent history of serious flooding, particularly in the shopping centre. Property was flooded in 1966, 1974, twice in 1979 (see Figure 2.5.3), and to a lesser extent in 1990. The town appears to be vulnerable to flooding for a number of reasons (see Figure 2.5.4):

- the narrowing of the floodplain at Blandford between the south and the north bank
- Blandford Bridge and its causeway which crosses the floodplain. In times of high river flows the Bridge presents an obstruction, which causes headwater to build up north of the causeway
- the floodplain downstream from the bridge is constricted for a number of reasons. The disused town swimming pool acts as a constriction on the north bank, and on the south bank there is high ground on which the Hall and Woodhouse Brewery is built
- the old railway embankment downstream from the town led to further constriction. This was removed with the completion of the by-pass in 1985

Figure 2.5.3 *Floods in Blandford, Dorset, 1979*

- groundwater from the surrounding hills usually flows through the gravel which underlies Blandford and into the river. When the river is high such flow is prevented, with the result that groundwater levels in Blandford rise, and cellars within the town are flooded
- highway drainage, and the drainage from the catchment of the Pimperne Brook (17 km^2) normally pass directly into the Stour, but in times of flood are unable to do so, and water will back up these systems, and add to the flooding effect.

As a result of the 1979 floods Hydraulics Research carried out a modelling study of the Stour, which included the necessary analysis for the design of works to protect Blandford. Two main stages of the work to protect Blandford have been completed. Stage One included the building of a flood bank on the north bank upstream of Blandford Bridge, and a floodbank and floodwall on the south bank downstream of Blandford Bridge (see Figure 2.5.4). Stage Two completes the flood defences of Blandford, and is shown in Figure 2.5.5. These defences are designed to protect 180 properties against a one in 200 year flood (the **return periods** for the May and December flood events of 1979 were one in

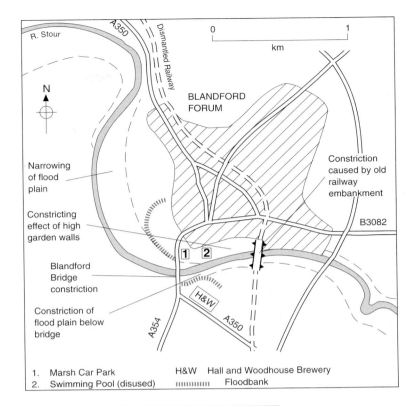

Figure 2.5.4   *Causes of flooding in Blandford Forum*

12 years and one in 65 years respectively). Floodwalls have been designed to blend

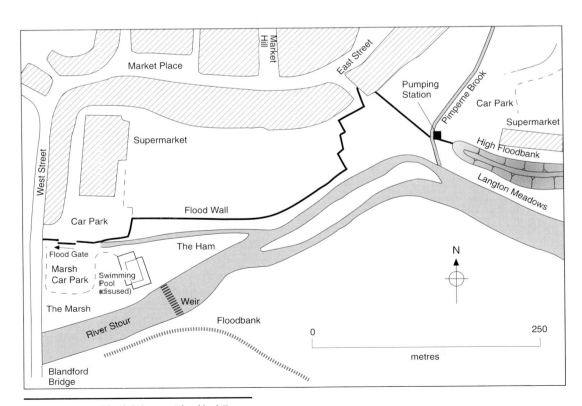

Figure 2.5.5   *Flood defences, Blandford Forum*

with the building materials used in the conservation area of Blandford. Particular attention has been given to the problem of floodwater accumulating behind the floodwall in times of heavy rainfall and high flow. This water will be channelled in a pipe that leads to the pumping station at Langton Meadows at the downstream end of the flood defence works, which will also handle flows from the Pimperne Brook. Benefits from the scheme amount to £1.64 million, and the overall cost £1.33 million, giving a cost/benefit ratio of 1.23.

STUDENT 2.7 ACTIVITY

1 Draw up a table showing the costs and benefits of the Blandford Forum Defence Scheme. You will need to distinguish between the **tangible** and **intangible** costs and benefits. Tangible costs and benefits are those to which we can assign a definite monetary value, e.g. costs incurred in the actual defence works themselves and benefits derived from damage averted to property and transport routes that would have been caused by floods had the defence works not been there. Intangible costs would include such things as damage to landscape, to flora and fauna, and other amenity damage. Benefits include avoidance of stress and worry to inhabitants in flood prone areas.

2 To what extent does this flood defence scheme incorporate the environmentally sensitive principles of river management discussed in section 2.1?

ESSAYS

1 Discuss the reasons why successful river management should adopt an integrated drainage basin approach.

2 Assess the management problems that arise from conflicting uses of rivers within a catchment.

FIELDWORK OPPORTUNITIES & PROJECT SUGGESTIONS

1 Select a small stream that is prone to flooding. Take series of cross-sectional measurements at different locations to show how hydraulic efficiency and discharge may vary. Use local records to map the area that is likely to be flooded by floods of different periodicity (i.e. 1 in 10 yeas, 1 in 50 years) – the Environmental Agency may have such data. Classify the land use types that are vulnerable to such flooding. Then map all of the flood defence measures that are visible, and use local records to see how effective they are. Alternatively, if there are no flood defence measures, then you could make some suggestions as to what management initiatives might be used.

2 Identify a local river that is intensively used for fishing. Assess the quality of the water in the river at a number of locations, measuring pH, dissolved oxygen, water temperature, and, if possible, test for a number of pollutants, and map your results. Contact the Environment Agency, whose catchment officer will be able to give you details of changing fish populations in the river, and management methods used to establish and maintain a healthy fishery. Local angling clubs may be contacted to see what informal management initiatives are used.

3

Managing Upland Environments

KEY IDEAS

- Upland environments in Britain and Ireland are sensitive and fragile, and need careful management
- Upland ecosystems are much valued for their diversity, and rarity, and require skilful conservation practice
- Upland environments are home to different groups of people, and their needs and aspirations need to be taken into account in the development of any management initiatives
- In many upland areas the needs of conservation, and the demand for recreation are likely to cause conflict
- Management of conflict needs to reflect different attitudes and values, and can often be resolved by sensible compromise

3.1 Introduction

The uplands of Britain and Ireland are remote and beautiful, yet often harsh and forbidding (Figure 3.1.1). The management of these lands is therefore inherently difficult. Uplands extend over between a third and a half of Britain and Ireland, yet carry only a tiny proportion of the population. The pattern of management is again a fragmented one, largely because of the pattern of ownership. Much of the land is in private ownership, which in areas like the Highlands of Scotland, and the uplands of western Ireland is undergoing important changes. The newspaper extract (Figure 3.1.2) indicates the nature of the change in Scotland.

STUDENT **3.1** ACTIVITY

1 Summarise the main features of ownership change that are taking place in the Scottish uplands
2 What are the main implications for upland management inherent in these changes?

Apart from land in private ownership, important areas of land are owned by bodies committed to a particular form of land use, such as forestry (the Forestry Commission), water resource development (Water Authorities) military training (the Ministry of Defence) or mineral resource exploitation. Indication of the conflicts that are likely to arise from different forms of land use are shown in Figure 3.1.3.

Figure 3.1.1 *Glen Affric, Scotland*

To many the uplands represent a sense of timelessness an retreat, an expansive, yet varied and intimate arena, where emotional and psychological revival can be encouraged by "getting away from it all".

Geoffrey Sinclair:
The Uplands
Landscape Study.

Who owns Scotland?

HALF of Scotland is owned by just 500 people, few of whom are actually Scots. As Britain's great land-owning aristocratic families decline, a new breed of foreign laird is exploiting Scotland's arcane land laws to buy up tracts of the Highlands and islands – Europe's last great wilderness.

The revelation comes in two new books which examine who owns Scotland.

They show that most lairds no longer hail from Britain's tweed-clad huntin', shootin' and fishin' classes; these days your local feudal overlord is more likely to be a self-made continental millionaire or an entrepreneur from Dubai, Egypt, Malaysia, Hong Kong or plain old America.

The findings have sparked a political row north of the border. Many of the new lairds are absentee land-owners who, environmentalists claim, neglect Scotland's greatest asset – the land itself.

Nationalist MPs and crofters, frustrated by the failure of Westminster politicians to bring Scotland into line with England and other European nations by abolishing feudal structures and regulating land use, are drawing up plans to limit foreign land ownership and introduce environmental codes for all estates. They want ministers to compile a full public Land Register.

The two books, *Who owns Scotland now?* and *Who owns Scotland*, update John McEwen's ground-breaking attempt to sketch Scotland's land-owning geography 30 years ago. His study revealed that ancient British families dominated the hills, straths, glens and islands, controlling lucrative salmon beats and deer stalking from the Borders to Barra.

Since the Fifties and Sixties, however, the decline of some of the most distinguished and notorious names in the Highlands – the clan chiefs of the Frasers of Lovat, the Sutherlands and the Wills tobacco family – has paved the way for new owners to take to the hills.

Andy Wightman, author of *Who owns Scotland*, which will be published in April, explains: "Some of the old landowners like the Duke of Buccleuch, the Duke of Atholl and Cameron of Lochiel have survived. Their old money is still good and some of their estates have expanded. But other families have fallen on hard times and a new group of landowners has stepped in swiftly to take their place. Many of these are from overseas and as they move in, a new pattern of land ownership is emerging."

All over Scotland there are now glens and peaks that are forever Swiss, Danish, Malaysian, Middle-Eastern and American. One year ago the whisky distilling MacDonald-Buchanan family sold off the Strath Conon estate in Ross-shire, which they had held for three generations. The new kilted monarch of the 50,000-acre glen is Kjeld Kirk-Christiansen, who runs the huge Danish Lego corporation.

Visitors to Queen's View in Glen Avon, where Queen Victoria used to look out on her Royal fiefdom, now look down on land owned by the mysterious businessman behind the Kuala Lumpur-based Andras conglomerate. He bought the 40,000-acre estate, once owned by the Wills family, for £6m last year.

"It's a dramatic change," said George Rosie, the veteran Scottish land-reform campaigner. "In the 19th-century, parliament passed an Act allowing foreigners to buy any property. As Britain was then the biggest rooster on the midden, the idea that any foreigner would be able to buy into Britain was risible. But now there are millions of wealthy foreigners and Scotland is ripe for the plucking. And plucked we have been."

Some of the new wave of overseas buyers enjoy good relations with locals and have earned environmentalists' praise for their land management practices. Paul van Vlissingen, the Dutch businessman whose "holiday cottage" is the eight-bedroom whitewashed Letterewe lodge on the banks of Loch Maree in Wester Ross, has helped to fund a new swimming pool and re-introduce native woodland on his 80,000-acre estate.

Other lairds, however, have been accused of barring access to walkers and neglecting the natural environment. "Mountain Closed" signs have recently appeared on estates north of Ullapool. In Perthshire, His Excellency Mahdi Mohammed Al Tajir, from the United Arab Emirates who owns the Blackford Estate, home of Highland Spring mineral water, has been accused of abandoning farms on the slopes of the Ochil Hills.

Nationalist politicians say Scotland's free market in land – it is one of the few countries in Europe which allows wealthy foreigners to buy up unlimited amounts of land with no questions asked – has created a "land lottery". While far-sighted landowners are welcome, the say new measures should be introduced to limit the size of their holdings and to remove those who neglect their land. Later this year, the Scottish National Party, which has set up an independent land commission, will unveil new proposals. Labour, too, has pledged to use a devolved Scottish parliament to introduce changes, and the Scottish Crofters' Commission is encouraging crofting communities to raise money to take over their marginal plots. Even Michael Forsyth, the Scottish Secretary, has jumped on the ox-cart, announcing plans to hand over millions of acres of government-owned crofting and forestry estates to smallholders. In Scotland, land is suddenly a political issue.

Dr James Hunter, a Skye-based environmental historian, said: "Land has moved up the agenda ever since the first crofters took over their land in Assynt 1992. That showed that land ownership patterns could change. Since then we have had controversies over the Knoydart estate and other Highland wilderness areas. The prospect of devolution – a Scottish parliament would address these long-standing grievances – has also concentrated minds."

Independent on Sunday 25 February 1996

Figure 3.1.2   *Who owns Scotland?*

Often this multiplicity of land use occurs within a section of the uplands that has been designated as a National Park (England, Wales, and the Republic of

Ireland see map Figure 3.1.4). National Park Authorities are charged with the overall management of the area within their jurisdiction, although, since, in England and Wales, they themselves own very little of the land within the Parks, their degree of control depends on a range of planning regulations for all forms of development. In the Republic of Ireland the land within the Parks is actually owned by the state, which, theoretically should make management policies easier to operate, but it has to be remembered that the Irish National Parks are very limited in extent. Scotland and Northern Ireland have no National Parks, although proposals for National Parks in Scotland have existed since 1945, and have recently received further attention (see Figure 3.1.4). Areas of Outstanding Natural Beauty are designated in the uplands of England, Wales and Northern Ireland, but, once again, overall management, tends to focus on the operation of planning controls on development.

Apart from the broad overall supervision that exists through the planning controls, bodies such as the Countryside Commission in England and Wales, and Scottish Natural Heritage have management responsibilities. The Countryside Commission combined with the Lake District Planning Board in establishing the Upland Management Experiment in the 1970s, and a similar scheme operated in the Snowdonia National Park in the same decade. The principal role of Scottish Natural Heritage may be defined in the following extract from its Corporate Plan:

'Although Scotland's natural heritage remains one of the most important and distinctive in Europe it has deteriorated markedly in living memory.

SNH's central aim is to stop and reverse this deterioration, principally by influencing the policies of the Government and its agencies and by enabling practical action by all whose activities affect the heritage, both to ensure its conservation, and sustain its very high value as a source of great pleasure and fulfilment to people.'

Its analysis of the factors influencing Scotland's mountain natural heritage is shown in Figure 3.1.5.

On a smaller scale, bodies such as the National Trust, and the National Trust for

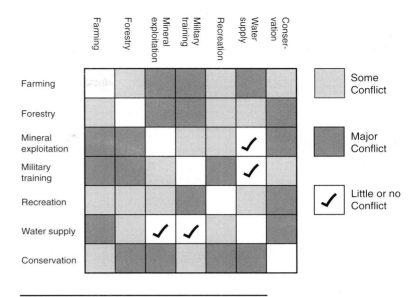

Figure 3.1.3 *Conflict matrix, upland environments*

Scotland have important management roles to perform in upland areas. Private bodies, with specific responsibilities such as forestry concerns, or broader management responsibilities, such as the large 'sporting estates' in the Highlands of Scotland, also have important roles in the overall management pattern of uplands.

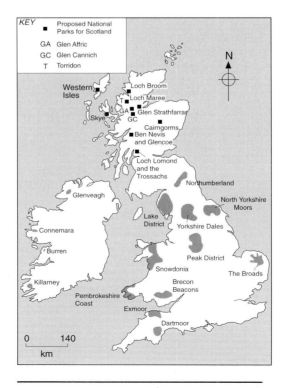

Figure 3.1.4 *National parks of Britain and Ireland and proposed for Scotland*

Factor	Impact trend	Influence	Location
Grazing sheep	Rising	Widespread conversions of moss and dwarf-shrub heath communities to grassland, and proliferation of local erosion.	South of the Scottish Highlands and locally in the central Highlands.
		Exacerbation of erosion on steep slopes.	Throughout Scotland, notably south of the central Highlands.
		Damage to fragile plant communities of springs and flushes, which can attract grazing sheep.	Throughout Scotland.
		The nests of ground-nesting birds such as dotterel are trampled.	Locally in parts of the central Highlands.
		May maintain diversity of some species-rich vegetation.	Central and south western Highlands.
Red deer	Rising	Suppression of scrub and woodland vegetation in open, high, sub-montane areas, and into the montane zone.	Throughout Scotland, but particularly in the central Highlands.
		Exacerbation of erosion on steep slopes and over peat hagg areas.	Throughout Scottish Highlands.
		The nests of ground-nesting birds such as dotterel are trampled by large herds when they occasionally stampede.	Central Highlands.
Acidic deposition	Rising	Direct deposition from clouds may favour the spread of grasses and sedges at the expense of globally rare mosses.	Problems are greatest in parts of Galloway and south of the Scottish Highlands.
		Increases in acidity of montane soils, increasing the prevalence of grasses.	Problems are greatest in parts of Galloway and south of the Scottish Highlands.
Recreation	Rising	Spread of footpaths and associated erosion, notably in steep or boggy areas.	Throughout Scotland, notably on Munro mountains.
		Intrusion in an otherwise wild landscape.	
		Presence of five down-hill ski centres (four expanding) which attract more people to the tops.	Throughout Scottish Highlands.
		Hill tracks used for transporting grouse and deer 'shooters' that attract hill-walkers.	Central and eastern Highlands.
Predators (crows and foxes)	Rising	There is a possible link between the increase in the numbers of predators observed on some tops with the increase in numbers of walkers. Further links are likely with moorland management lower down the mountainside (grouse moors with considerable predator control have fewer predators).	Throughout Scottish Highlands.

Figure 3.1.5 *Factors affecting Scotland's mountain heritage*

3.2 Case Studies of Upland Management

The case studies selected for examination in this section aim to show something of the scale and variety of upland management. The first study examines management of Glencoe, a major National Trust location in Scotland. Although it is regarded as a historic site of international importance, it is also a key site for British mountaineering. Furthermore, its unique geology has a high conservation value.

Limestone uplands are common to both Britain, and the Republic of Ireland, and short studies of management problems in the Yorkshire Dales and the Mendips indicate the range of problems and conflicts that have to be solved. Finally an overview of the management of the four National Parks of the Republic of Ireland completes the range of scales that are examined.

3.3 Managing Glencoe: the National Trust for Scotland

Charles Dickens and Dorothy Wordsworth present very different views of Glencoe. Dickens presents the Victorian view of Glencoe as a wilderness to be feared and loathed. Dorothy Wordsworth, conscious perhaps of her own Lake District mountains, accepted the mountain environment as something to be valued and cherished.

Glencoe (Figures 3.3.1 and 3.3.4) possesses some of Scotland's finest mountain scenery. The Glen is a deep glacial trough, overdeepened, and thus giving prominence to the great sweeps of mountain that form its northern and southern walls. To the north the sides of the trough rise up through scree fans and faces of bare rock to the precipitous ridge of Aonach Eagach, linking the two summits of Stob Coire Leith and Meall Dearg. On the south the magnificent peak of Buchaille Etive Mor (Figure 3.3.2) is first encountered after crossing Rannoch Moor, and this is then succeeded by a series of great buttresses and spurs that form the Three Sisters, rising to the summits of Bidean nam Bian and Stob Coire Sgreamhach to the south. The map (Figure 3.3.4) shows the extent of the National Trust for Scotland Property in

Glencoe, which has been acquired in a series of purchases from 1935 to 1993. Two major SSSIs are included within the National Trust Property, and the whole of the Glencoe Property is designated as a National Scenic Area.

The National Trust Property is of national and international importance for the following reasons:

- its outstanding geological and geomorphological features. In particular it is one of the finest examples of **cauldron subsidence** in the world (see

Figure 3.3.1 *Glencoe, looking east*

Glencoe itself is perfectly terrible. The Pass is an awful place . . . The very recollection makes me shudder.

Charles Dickens during his travels in Scotland.

The impression was, as we advanced up to the head of this first reach, as if the glen was nothing. Its loneliness and retirement made up no part of my feeling: the mountains were all in all.

Dorothy Wordsworth: Journals.

Figure 3.3.2   *Buachaille Etive Mhor*

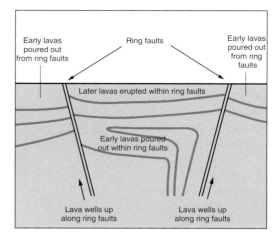

Figure 3.3.3   *Cauldron subsidence in Glencoe*

Figure 3.3.3) and has been an important field site and training location for geologists for well over a hundred years. Its mountain landscape displays all of the classic forms of glacial erosion, and glacial and **fluvioglacial deposition**;

- its SSSIs are of international ecological importance for the range of habitats which include grassland, broad-leaved and coniferous woodlands. These habitats shelter a range of rare birds and plants, particularly **bryophytes**;

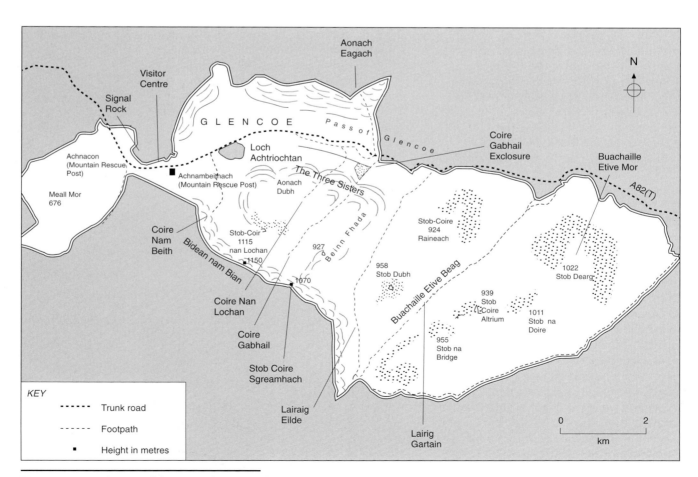

Figure 3.3.4   *Glencoe and the National Trust site*

- it is one of the classic centres of British mountaineering. It possesses nine Munros (summits over 3000 feet), 18 Tops (subsidiary summits over 3000 feet) and 20 separate recognised climbing crags. Many of the routes on the mountains were first explored by pioneering Victorian mountaineers, and continue to attract mountaineers particularly for the snow and ice-climbing in the winter. The number of mountaineering visits in a year is put at between 50 000 and 80 000;
- it is one of the major historic sites in Scotland, best known perhaps for the massacre of the Macdonalds by the Campbells in 1692.

MANAGEMENT ISSUES

The main management issues appear to be:

- maintaining and enhancing the ecological status of habitats within the property;
- access to the property: the problems associated with the A82;
- visitor management: maintaining the balance between conservation and recreation within Glencoe;
- camping: 'wild camping' outside of the formal campsites in the Glen;
- path management: increasing pressure on footpaths in the area is leading to serious erosion problems;
- mountain safety: as the area becomes increasingly popular with visiting mountaineers, the risk of accidents becomes greater, particularly in winter.

CONSERVATION AND ECOLOGY

Several of the small woodlands in Glencoe are valuable, since they are areas of relict native woodland: birch/rowan on ungrazed ledges, birch ash on Meall Mor, alder near Achnacon, willow scrub around Lake Achtriochtan and Scots Pine in Glen Etive. Natural regeneration is largely prevented by deer and sheep grazing. Active deer management (present total 126) is carried out to ensure a healthy balance between population and available grazing. An experimental enclosure was created at the foot of Coire Gabhail in 1984. This has led to widespread natural regeneration of birch rowan, willow and heather, much improving the site as a wildlife habitat. Overgrazing has contributed to large scale erosion on many of the valley slopes and loss of stability has led to landslips.

ACCESS TO GLENCOE

The A82 trunk route from Glasgow to Fort William is the key to visitor numbers in Glencoe. Heavy commercial traffic uses the A82 throughout the year, and this is compounded by intense tourist activity from Easter until late summer. Although extensive improvements to the route of the A82 have taken place, many of the bends are still sharp, and car parks and lay-bys along the road are dangerous and obtrusive, and often badly surfaced. The road is regularly blocked by debris slides and washouts, and engineering works would be needed to remedy this, but in such a physically dominant landscape they would have to be designed very sensitively.

VISITOR MANAGEMENT

Visitor management is focused on the Visitor Centre (Figure 3.3.6), which was built by the Countryside Commission for Scotland from 1974–6. It is still owned by Scottish Natural Heritage, but is currently managed by the Trust. The centre currently plays a major interpretive role in Glencoe, and provides nearly 100 car parking spaces. Originally it was designed to handle 50 000 visitors a year, but now receives nearly four times that amount. The centre now becomes extremely crowded during peak times in the summer, and a major management issue is now its relocation to another site.

CAMPING

No formal campsites exist within the Trust's property, although three are found just outside. This lack of facilities has led to much informal camping in the Glen, particularly at lay-bys. Although attempts to discourage such camping were very largely successful, they only succeeded in concentrating the camping adjacent to Clachaig Inn. Not only has this caused a public health hazard, due to the lack of toilet facilities, but it has also resulted in dangerous and illegal parking and has created visual intrusion. Similarly wild camping in Glen Etive has become obtrusive from time to time.

PATH MANAGEMENT

With increased visitor numbers to Glencoe, there is evidence of serious damage to the most heavily used paths in the Glen. Figure 3.3.4 shows the main path network. Pathcraft Limited, a subsidiary of Scottish Conservation Projects has undertaken various projects in Glencoe, with most work being concentrated on the Coire Gabhail path.

MOUNTAIN SAFETY

Safety of climbers and walkers has acquired a high level of awareness in Glencoe, since the area records the highest number of fatalities and injuries in any National Trust property in Scotland. This is principally because it is more heavily used than any other locations.

A NEW SITE FOR THE VISITOR CENTRE IN GLENCOE

Since the Visitor Centre was built in 1974–76, numbers of visitors have grown considerably, and it is no longer able to cope with the increased use it receives. It is operating at four times its designed capacity, and in the peak season it is frequently overcrowded. Working conditions for staff are poor, and it is suffering from structural problems. The centre is really too small, since storage space is extremely limited and there is no provision for educational groups or the ranger service. Now it is necessary either to extend and repair the existing property, or to seek to relocate it. Apart from the option of extending the present building, two options exist at the Forestry Enterprise site at Leacantuim (see Figure 3.3.5).

THE EXISTING SITE
Advantages
- it possesses superb views of Glencoe
- it occupies an attractive setting close to the river, with the footbridge over the falls and the picnic site beside the river
- it is close to the walks at An Torr and Signal Rock

Disadvantages
- it is an extremely restricted site with little opportunity for expansion
- its sensitive location at the 'core' of Glencoe
- the visual impact emanating from the unscreened car park

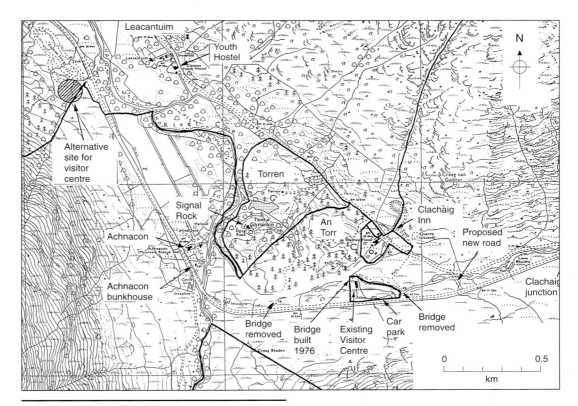

Figure 3.3.5 *Clachaig and Visitor Centre location plan*

Figure 3.3.6 *Glencoe Visitor Centre*

THE FORESTRY ENTERPRISE CAMPING AND CARAVANNING SITE
(Figures 3.3.7 and 3.3.8)

Advantages
- it is visually associated with the lower glen
- it is largely screened by mature woodland
- it offers attractive views of Glencoe
- it possesses good interpretive opportunities
- it possesses space for expansion

Disadvantages
- it is located away from the essentially wild part of Glencoe, and the wilderness aspect is lost
- it is located slightly away from the A82
- although it is screened from most low-level viewpoints, it is readily seen from major ridges and hilltops

The Options
Option 1: renovation and extension of the present visitor centre: Cost £0.8–£0.9 million.
Option 2: building a new centre at the existing site: Cost £0.9–£1.2 million.
Option 3: building a new centre at the Forestry Enterprise site in Lower Glencoe: Cost £1.1–£1.3 million. Within this option there are two choices
  i acquisition of part of the site, with a contribution towards upgrading facilities
 ii acquisition of the entire site, with the National Trust for Scotland managing the campsite.
1 Using the maps, and the text, discuss the merits of the three options.

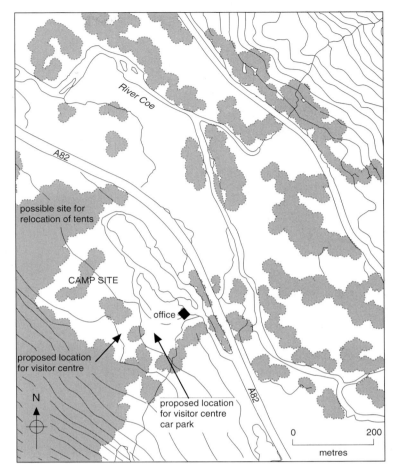

Figure 3.3.7 *Forestry Enterprise campsite*

2 Explain why different groups of people might have different ideas on the best option
3 Make a choice of the best option, giving your reasons for your selection.

Figure 3.3.8 *Forestry Enterprise campsite, Leacantuim, Glencoe*

3.4 The Management of Limestone Uplands

In Britain and Ireland, there are a number of extensive areas that are underlain by Carboniferous Limestone (Figure 3.4.1) (significantly, perhaps, originally known as Mountain Limestone in Britain).

In parts of the Pennines, around the fringes of the South Wales coalfield and the margins of the Lake District, this limestone forms characteristic **karst** scenery.

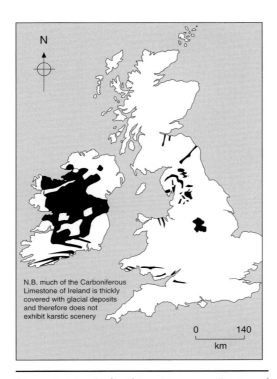

N.B. much of the Carboniferous Limestone of Ireland is thickly covered with glacial deposits and therefore does not exhibit karstic scenery

0   140
km

Figure 3.4.1 *Carboniferous Limestones, Britain and Ireland*

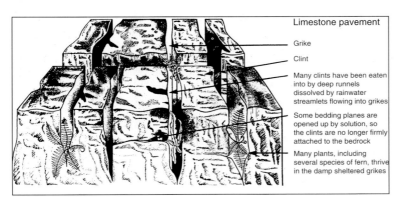

Limestone pavement

Grike

Clint

Many clints have been eaten into by deep runnels dissolved by rainwater streamlets flowing into grikes

Some bedding planes are opened up by solution, so the clints are no longer firmly attached to the bedrock

Many plants, including several species of fern, thrive in the damp sheltered grikes

Figure 3.4.2 *Limestone pavement*

Although considerable areas of Ireland are underlain by Carboniferous Limestone, much of it is masked by glacial deposits, and the most distinctive karstic features are found in the Burren of County Clare. One of the most distinctive features of karst scenery is the limestone pavement. The first section of this consideration of limestone uplands examines some of the problems of managing limestone pavements.

Not all limestone uplands in Britain exhibit pavements, with the Mendips in southern England being an example. The Mendips are built of Carboniferous Limestone, which is a particularly important mineral resource. Since the Mendips are the only source of such limestone in the south, it means that the limestone is much in demand and is intensively exploited. In the Mendips, the main management issue is the control of the exploitation of the limestone as a mineral resource, and this forms the theme of the second section.

THE MANAGEMENT AND CONSERVATION OF LIMESTONE PAVEMENTS

Limestone pavements are areas of bare limestone, that are broken into blocks (clints) by a series of deep fissures (grikes) that have been weathered out along joints (see Figure 3.4.2). The best examples of pavements are found in the Ingleborough district of Yorkshire, and in the Burren, although patches occur in South Wales, around the edge of the Lake District, and on Cambrian limestone in Skye and in Sutherland. All of these limestone areas have been extensively scoured by glaciation, and this appears to be a pre-requisite for their development. Glacial scouring has been responsible for the removal of loose weathered debris on the surface of the limestone, thus exposing the bare pavements. Additionally, the limestone needs to be pure and well bedded if it is to form well developed

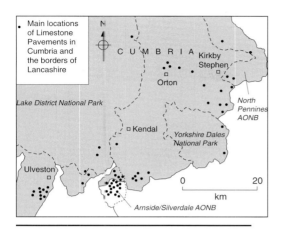

Figure 3.4.3 *Threatened limestone pavements in south east Cumbria and Lancashire*

pavements. The pavements vary enormously in form, from those with deep grikes to those where the clints are separated by wide vegetated furrows, no doubt the result of different erosion history and **lithology**.

Apart from their interest to the geomorphologist, pavements are much valued for their ecological significance. Within the grikes, shade and humidity encourage the growth of a distinctive plant association, including such woodland plants as hart's tongue fern, herb robert and dog mercury. Plants of rocky habitats are also found on upland pavements. Fauna include a wide variety of insects, although the only species of bird found in the grikes is the wren.

Limestone pavements have been seen throughout history as a source of building stone, and for stone wall construction. In the past stone was removed with care, but now tractors with haulage attachments can break up massive pavements, and explosives are actually used in some places. Within Britain there are only 2150 hectares of limestone pavement remaining and in areas like Cumbria almost half has been damaged (Figure 3.4.3).

The rockery trade sees limestone pavements as a valuable resource (Figure 3.4.4), and a conservation/resource conflict is evident between the two interests. Under the Wildlife and Countryside Act of 1981, there is provision to make Limestone Pavement Orders in respect of pavements of particular value. Orders are legal means of preventing the removal or damage of limestone in specially designated areas. Limestone Pavement Orders do not affect a farmer's right to graze stock over an area. Furthermore, the farmer is allowed to come to some agreement with the authority serving the order as to the amount of loose stone that he can remove for the repair of walls.

STUDENT ACTIVITY

1 Why are Limestone Pavement Orders necessary?
2 Discuss some of the difficulties of implementing the Orders.

MANAGING LIMESTONE AS A MINERAL RESOURCE: THE CASE OF THE MENDIPS

The principal uses of limestone as a mineral resource are shown in Figure 3.4.5. Most of the limestone that is extracted in

Threat to rare stone landscape

A RARE Yorkshire Dales rock formation, formed by thousands of years of wind and rain, is under threat because the National Park is powerless to prevent a farmer removing the stone, *writes Oliver Gillie*.

Winskill Stones are a geological formation known as limestone pavement and protected under the Wildlife and Countryside Act because it is a rare landform and provides an unusual habitat for wild plants, but Alec Robinson has a licence to take the stone that is valid until 2042.

When the pavement is broken up the naturally sculpted stones fall into pieces highly prized by gardeners and sell for up to £140 per tonne. However, a pressure group, the Limestone Pavement Action Group, wants to make gardeners ashamed of using this stone and hope to force it off the market.

Clive Kirkbride, landscape conservation officer with the national park, said: "This stone has been used for centuries for building and most of the sites have been damaged. There is relatively little left. Limestone pavement is now recognised under the EC habitats directive as a rare landform and a way must be found to revoke old mining concessions as soon as possible." The park wants to buy the 70 acres covered by Mr Robinson's concession but not at a price he can accept.

Figure 3.4.4 *Threat to Rare Stone landscape*

Use	Consumption (thousand tonnes)
Iron and steel	2872
Soda ash (sodium carbonate)	990
Sugar refining	245
Glass making	325
Precipitated calcium carbonate	100
Other chemical uses	173
Animal feeds	300
Agriculture	1950
Sea water magnesia	142
Paper filler and coating	305
Rubber and plastics	155
Other pigments	340
Total	7897

Figure 3.4.5   Principal uses of limestone

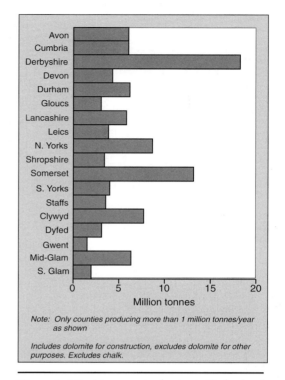

Note: Only counties producing more than 1 million tonnes/year as shown

Includes dolomite for construction, excludes dolomite for other purposes. Excludes chalk.

Figure 3.4.6   Limestone production in England and Wales

Figure 3.4.7   The Mendip Hills

the 14 quarries in the Mendips is for constructional uses. Only in a few quarries is chemical grade limestone (very pure and of high quality) produced. The Mendips rank second after Derbyshire in the amount of limestone produced (see Figure 3.4.6). Environmental conflict is inevitable when important limestone resources occur in an Area of Outstanding Natural Beauty (see Figure 3.4.7). The dilemma is well illustrated in Figure 3.4.8.

Possible values of extracted quarry rocks								
	Hard limestone		Granite		Basalt		Gritstone	
Cost of extraction	Low	5	High	2	High	2	V. high	1
Common aggregate	Good	5	Good	5	Good	5	Good	5
Base aggregate	V. good	5	Good	4	V. good	5	Good	4
Wearing course	Not used	0	Medium	3	Good	4	V. good	5
Quarry wastage	Low	4	Medium	3	High	2	V. high	1
Building stone	Used	2	Useful	3	Rarely used	1	Used	2
Chemical	Yes	5	No	0	No	0	No	0
Extracted stone (total score)		26		20		19		18

Score: 1 = poor, 5 = good

Relative values of quarry rocks when left in the ground							
	Hard limestone		Granite		Basalt		Gritstone	
Aquifer	Good	4	Poor	1	Poor	1	Occasional	2
Landscape	V. good	5	V. good	5	Good	4	V. good	5
Attractions (examples in the British Isles)	Cheddar Gorge Bath Hot Springs Wookey Hole Caves Malham Cave and Burren Pavements	5	Dartmoor tors Land's End and Lundy Island	4	Giant's Causeway Fingal's Cave and High Force	4	Pen-y-Fan Croagh-Patrick and Suilven	4
Speleology	Yes	5	No	0	No	0	No	0
Walking, climbing	V. good	5	V. good	5	Good	4	V. good	5
Open space	Yes	5	Yes	5	Yes	4	Yes	5
Paleolithic remains	Many	5	Many	5	Few	1	Few	1
Mines, minerals	Many	5	Many	5	Few	1	Few	1
Agricultural soil	Good	4	Poor	2	V. good	5	Poor	2
Fauna and flora	V. diverse	5	Limited	2	Good	4	Limited	2
In situ stone (total score)		48		29		27		26

Figure 3.4.8   Limestone, the environmental dilemma

STUDENT   3.3   ACTIVITY

1 Explain why these tables illustrate the dilemma facing planners responsible for resource development and conservation.
2 Is there any realistic resolution of the dilemma?

The principal environmental losses to the Mendips are summarised in Figure 3.4.9. An ingenious way of assessing the conflicting values of exploiting and conserving the Mendips is to calculate what is known as the Mendip limestone budget. Quarried limestone constitutes the debit account, and in 1992 its value amounted to £46 million. The credit account relates to in situ limestone, and is more difficult to calculate:

Tourist income generated by Mendip limestone (limestone dependent attractions in Mendips)   £36 million
Hot Springs at Bath supplied by the Mendips (assume 10 per cent of annual tourist income in Bath)   £12 million

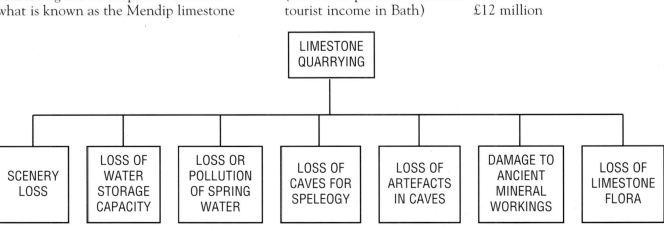

Figure 3.4.9   Environmental losses in the Mendips as a result of limestone quarrying

Water supply from the
Mendips
(cost of alternative source
= R. Severn)   £24 million
Income from cave explorers
and rock climbers   £1 million
Total   £73 million

Intangible matters, such as the value of the
Area of Outstanding Natural Beauty of the
Mendips, and the income generated from
foreign tourists complete the credit
account. Intangibles associated with the
limestone quarrying are mostly negative
such as noise, traffic hazards, dust in dry
weather, mud in wet weather, devaluation
of property and loss of open space.

Thus on an annual cash flow basis the
value of quarried Mendip limestone (£46
million, the annual limestone budget), less
intangibles and £3 million road charges is
very much less than the value of in situ
Mendip limestone at £73 million plus
intangibles.

In the longer term (the total limestone
budget), the Mendip limestone resource, if
it were extracted at the 1992 rate (£46
million) would last for 5230 years. At the
natural wastage rate (through erosion – 28

000 tonnes a year), the Mendips would last
2.4 million years. Put this way the Mendips
are worth £240 billion as limestone
aggregate, and £175 trillion as limestone
hills. Although the arithmetic may be
questioned the calculations have a very
important environmental and conservation
message.

In reality management of limestone as a
resource for an area such as the Mendips is
vested in the local authority. The planning
system is shown in Figure 3.4.10. The role
of the planner is summed up as follows:
'The task of the planner is to prepare
advice for politicians, . . . which they can
accept, modify or reject. In preparing this
advice they must consult, listen and
balance the arguments. The essence of
planning judgement lies in determining the
balance of advantage between the need for
development and the amount of
environmental damage it may cause.'

STUDENT   **3.4**   ACTIVITY

1 How realistic is the idea of
  a an annual limestone budget
  b a total limestone budget

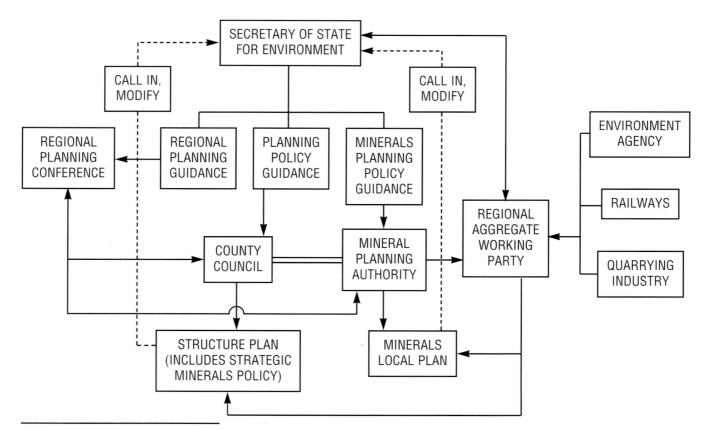

Figure 3.4.10   *Minerals planning system*

3.5 Managing National Parks in Ireland's Western Uplands

The four National Parks in the Republic of Ireland are shown in Figure 3.1.4. It will be seen from the map that all four of the National Parks are in the upland west of Ireland, Killarney in County Kerry, the Burren in County Clare, Connemara in County Galway, and Glenveagh in County Donegal. National Parks are managed by the National Parks and Monuments Branch of the Office of Public Works. The land within the National Parks has been acquired by the state, either through gift, bequest or purchase, unlike the British National Parks, in which much of the land is privately owned. National Parks, in the Irish context, are defined as 'areas which exist to conserve natural plant and animal communities and scenic landscapes which are both extensive and of national importance, and under conditions compatible with that purpose, to enable the public to visit and appreciate them.' Five basic objectives for National Parks are:

- to conserve nature within the Park;
- to conserve other significant features and qualities within the Park;
- to encourage public appreciation of the heritage within the Park and the need for conservation;
- to develop a harmonious relationship between the Park and the community;
- to enable the Park to contribute to science through environmental monitoring and research.

Of these, nature conservation takes precedence over the others should any conflict arise.

KILLARNEY NATIONAL PARK

Killarney National Park is the largest of the three Irish National Parks (just over 10 000 hectares, established in 1932) compared to Connemara (2669 hectares, established in 1980), Glenveagh (9667 hectares, established in 1975) and The Burren, 1128 hectares (established in 1991). The original park was given to the

Irish nation in 1932 as the Bourne Vincent Memorial Park, or the Muckross Estate, and was managed more or less as a large farm, both before and after the state had acquired the property. Later acquisitions have added both land and water to the property. It has a high scenic value, centred on the three lakes, and its ecology is of international significance, with its oakwoods forming the largest remaining remnant of the original oakwood cover of Ireland. A range of cultural resources, covering a wide span of Irish history is also an important feature of the Park.

The principal spatial organisation of management is manifested in the creation of zones within the Park (Figure 3.5.1). Four zones are distinguished:

- the natural zone, which possesses thirteen areas or features notable for

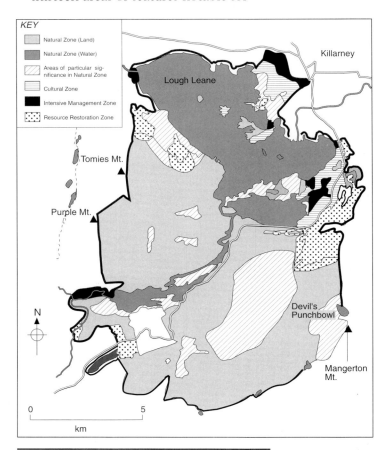

KEY
- Natural Zone (Land)
- Natural Zone (Water)
- Areas of particular significance in Natural Zone
- Cultural Zone
- Intensive Management Zone
- Resource Restoration Zone

Killarney
Lough Leane
Tomies Mt.
Purple Mt.
Devil's Punchbowl
Mangerton Mt.
N
0   5
km

Figure 3.5.1   *Zoning system, Killarney national park*

their ecological or geological significance. These include woodlands, bogs, wintering areas for migratory geese and red deer breeding grounds. Nature conservation is the primary objective in this zone;

- the cultural zone, which possesses artificially improved estate land, with woodland and grassland, and also a range of significant archaeological and historic sites. Conservation of the human legacy is the main aim here;
- the intensive management zone, which includes areas used intensively for visitor services, such as Muckross House and Gardens and Park administration;
- the resource restoration zone, which comprises plantations of conifers and other non-native trees, which was managed as a commercial timber zone. Future management will aim at restoring the plantations to a state which conforms to the general aesthetic aims of the Park.

THE BURREN NATIONAL PARK

The Burren is a wild and desolate karstic upland in County Clare. It has long attracted attention from visitors, not only

Figure 3.5.2 *Limestone Pavement, the Burren, County Clare*

on account of its distinctive scenery (see Figure 3.5.2) but also because of its flora and fauna, its abundant archaeological remains, and its abiding historical interest. At present, the area of the National Park is relatively small, 1128 hectares, although it is planned that the eventual size will be approximately 3000 hectares. It was designated in 1991, and plans for a new visitor centre were announced at the same time, but were the subject of considerable controversy. Opponents of the scheme took the Office of Public Works to court and won an injunction that forced the Office of Public Works to go through normal planning procedures.

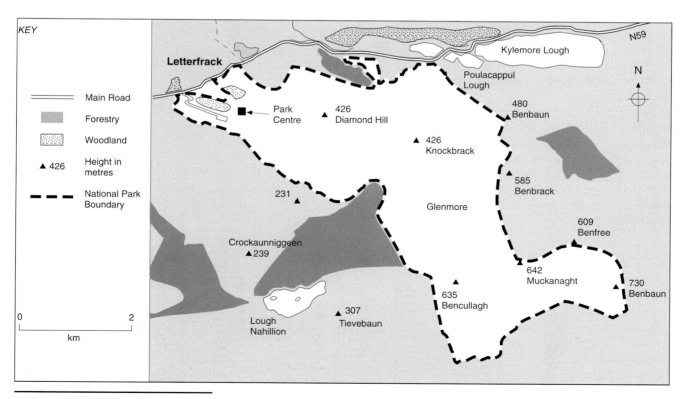

Figure 3.5.3 *Connemara National Park*

CONNEMARA NATIONAL PARK

Connemara National Park (Figure 3.5.3) is much smaller than Killarney, but presents different management issues. The centre of the Park is focused on the wild and rugged mountains of Benna Beola, or the Twelve Pins of Connemara. These mountains (Figure 3.5.4), only some of which are included within the limits of the Park, are formed mainly of PreCambrian quartzites, with some **schists** and marbles outcropping round the margins. Within the Park, Muckanacht (see Figure 3.5.3) is the only peak formed of schist, all of the others are of quartzite. All of the area has been intensively glaciated. During the **Pleistocene Period**, Connemara was covered with an ice dome some 1000 m thick, from which glaciers flowed out radially. They cut deep glacial troughs, two of which delimit the Connemara National Park on the north and south. Apart from these troughs the Pleistocene glaciation has left behind a rich legacy of glacial and fluvioglacial landforms within the confines of the Park.

Connemara is particularly important from the standpoint of natural history (Figure 3.5.5). Although the main areas of **blanket bog** are found in that part of Connemara which lies to the south of the National Park, there are important remnants within its boundaries. These bogs consist of compacted layers of waterlogged peat, which carry a surface layer of living vegetation. The bogs are particularly diverse ecologically, with 'a rich mosaic of hummocks and hollows, sedge lawns, flushes and bog pools'. Rising above the blanket bog remnants are the Twelve Pins, and their distinctive montane plant communities growing on the thin quartzitic soils. These arctic-alpine communities are rich in flowering plants, lichens, mosses and liverworts.

Much of the Connemara National Park was originally part of the Kylemore Abbey Estate and the Letterfrack Industrial School. Kylemore Abbey itself lies outside of the Park, but the buildings of the Letterfrack Industrial School are used today as part of the visitor centre and the administrative headquarters of the Park. In a more remote situation than Killarney, and consisting principally of high

Figure 3.5.4   *Peaks in the Twelve Bens, Connemara, County Galway*

mountains and blanket bog, Connemara receives fewer visitors.

STUDENT   3.5   ACTIVITY

1 Discuss the ways in which management issues in Connemara differ from those in Killarney.
2 Use Figure 3.5.3 to suggest possible zoning of the Connemara National Park, based on the Killarney model. If you do not think zoning is appropriate, explain your reasons.

THE GLENVEAGH NATIONAL PARK

Glenveagh National Park is in County Donegal, and lies astride the Derryveagh Mountains. The summits of the two highest mountains in County Donegal, Errigal (Figure 3.5.6) and Slieve Snacht both lie within the Park. Like both Connemara, and Killarney, the physical

Figure 3.5.5   *Peat Bog, Connemara, County Galway*

Figure 3.5.6   *Errigal, Glenveagh National Park, County Donegal*

landscape of Glenveagh bears the imprint of intense glaciation, with such fine troughs as the Poisoned Glen and the valley containing Loch Veagh.
Oak and birch woodlands in the valleys provide reminders of Killarney, whilst extensive tracts of bog and moorland suggest similarities with Connemara. Two large herds of red deer (not of Irish stock)

are the main feature of the wildlife. As is the case with Connemara, Glenveagh is a wilderness landscape, with the only human activity being that focused on Glenveagh Castle and gardens and the purpose built Visitor Centre. Glenveagh estate was purchased by the state in 1975, and the Castle and gardens were presented to the Irish state in 1981.

STUDENT   **3.6**   ACTIVITY

1 Suggest why zoning might not be appropriate for the management of the Glenveagh National Park.
2 Attempt a comparison of the management issues and needs of the Killarney, Connemara and Glenveagh National Parks, and explain any important differences that you can identify.
3 Why should the Irish National Parks present fewer management problems than those in England and Wales?

ESSAYS

1 Examine how conflicts between conservation and recreation in upland areas can best be managed.

2 Explain why upland environments are particularly fragile, and need sensitive management.

FIELDWORK OPPORTUNITIES & PROJECT SUGGESTIONS

1 Examine the way in which a particular footpath, or network of footpaths have been managed in an upland valley bottom, and its enclosing slopes. Map the path, or network, in a scale of 1 : 25 000. In the field survey, classify the surface of the footpaths, according to the material used, and then map the distribution of the different materials used. Using a numerical scale, assess the effectiveness of the different materials used. Map the distribution of stiles and gates along the path or its network, and assess their effectiveness (safety, ease of use). You could extend this project by introducing a user survey.
2 Compare the upper and lower parts of an upland valley, and the way that they are managed. A useful way to do this is

to construct a cross profile of the valley, either by using a large scale map (1 : 10 500) or, if the valley is manageable enough, by using a clinometer, surveying poles, and measuring tapes. Plot the land on your cross section, plot the field boundaries (walls, fences, hedges), and annotate the cross section to show their current state. Locate old abandoned farmbuildings, or sheep shelters in the upper valley section, and note all fields that are no longer managed, and that have been invaded by bracken. In the lower valley section, locate the field boundaries, note the quality of pasture, stocking levels of animals. NB It will be essential and courteous to enlist the aid of the relevant landowner in both parts of the valley.

Managing Physical Resources

4

KEY IDEAS

- Small countries like Britain and Ireland will seek to make maximum use of their physical resources
- Physical resources are often located in areas with fragile ecosystems and valued landscapes
- Decisions over exploitation of resources have to balance the national interest against local and regional concern over environmental damage
- Skilful and sensitive management is necessary to ensure that resource development is in harmony with the surrounding natural and human environment

4.1 Introduction

Physical resources include those that are finite, such as mineral resources and fossil fuels, and those that are recyclable, such as water and waste. Because Britain and Ireland are relatively small countries and dependent on overseas suppliers for an important range of physical resources, such as a whole range of mineral ores, they need to maximise the use of all of their indigenous resources. Management of these resources is increasingly concerned with achieving a balance between the utilisation of the resource and minimising the environmental impact resulting from its exploitation. Figure 4.1.1 shows some of the potential environmental impacts of the utilisation of physical resources. As Judith Rees points out 'before the emergence of the environmental movement as a political force, the level of interest among social scientists in resource and environmental questions was extremely limited'. Figure 4.1.2 shows the growth in membership of British Environmental organisations.

STUDENT ● 4.1 ACTIVITY

Study Figures 4.1.3 and 4.1.4 which show the most important environmental problems in England, Wales and Scotland, and the degree of concern over environmental issues in those countries. For each of the issues identified, consider the extent to which the utilisation of physical resources, as defined at the beginning of the chapter, is responsible, or partly responsible for the problem. Summarise the results of your analysis in a table.

Central government in both the United Kingdom and the Irish Republic is responsible for the planning regulations within which the development of physical resources takes place, although the granting of planning permission is usually delegated to local government. One of the most important features of the

. . . motorways, airports, power stations and reservoirs are more or less permanent features and [their development] must be balanced against the destruction of the corresponding amenity. But, in the case of exploiting mineral resources, it is a matter of balancing a permanent destruction of amenity against a short term material gain.

Peter J Smith
The Politics of
Physical Resources.

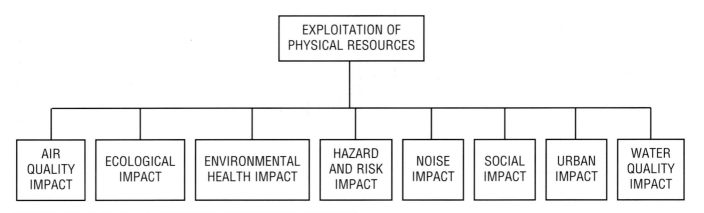

Figure 4.1.1   *Environmental impact of exploitation of physical resources*

management of physical resources in Britain over the last 15 years has been the move from the public to the private sector. Thus all energy and water resources are now effectively managed by the private sector, although the Government has regulatory controls in place over such matters as environmental standards and pricing. In this respect bodies such as the new Environmental Agency are responsible for standards not only within the water industry, but in the whole range of industries that are involved in the development of physical resources.

In the Irish Republic state-sponsored bodies such as the Electricity Supply Board, and the Gas Supply Board are responsible for the supply of electricity and gas, and the overall management of energy resources. Bord na Mona is another State company responsible for the management of the peat resources of the Irish Republic. Responsibility for water resources in the Irish Republic is in the hands of local

authorities, although the Department of the Environment has theoretical overall control.

In both the United Kingdom and the Irish Republic European legislation is playing an increasingly important part in the management of physical resources. One of the most significant pieces of legislation affecting the management of physical resources is Directive 85/337. This requires that environmental implications of major projects be considered at the earliest possible stage and that an environmental assessment, which includes an environmental baseline study, and an impact statement, has to be undertaken for all projects that are likely to produce significant environmental effects. In the Irish Republic this Directive has been incorporated into Irish law, and all mining projects now require an environmental impact statement by a person or company approved by the Minister of Energy.

	in thousands		
	1971	**1981**	**1990**
National Trust	278	1046	2032
National Trust for Scotland	37	110	218
Royal Society for the Protection of Birds	98	441	844
Greenpeace UK	–	30	372
Civic Trust	214	–	302
Royal Society for Nature 'Conservation'	64	143	250
World Wide Fund for Nature	12	60	247
Friends of the Earth	1	18	110
Ramblers Association	22	37	81
Woodland Trust	–	20	66
Council for the Protection of Rural England	21	29	44

Figure 4.1.2   *Growth in membership of voluntary environmental organisations 1971–90*

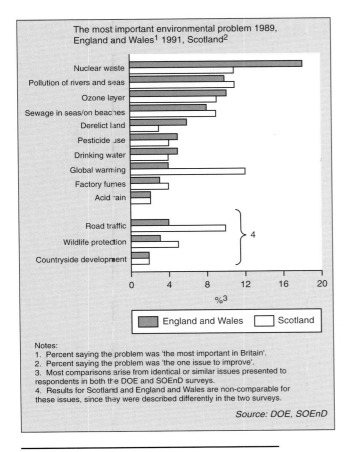

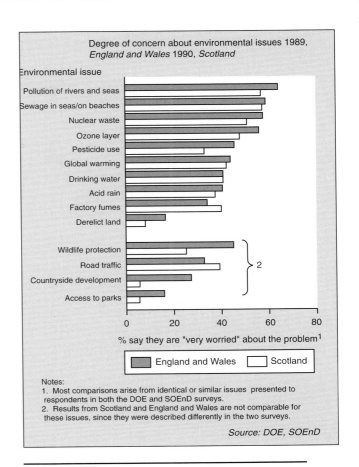

Figure 4.1.3 *Most important environmental problems*

Figure 4.1.4 *Degree of concern about environmental issues*

Case Studies in the Management of Physical Resources

4.2

These case studies are arranged at a variety of scales to examine the different aspects of management at these levels. In the Isle of Purbeck in south Dorset several important mineral resources are concentrated within a relatively small area. In the heathlands, oil, gas and ball clay are resources of national, and in some cases, international importance. To the north of the heathlands the terraces along the Rivers Piddle and Frome are worked for aggregate supplies for a considerable area in central southern England, whilst the stone industry in the south of Purbeck supplies a

similar area. Management of these resources is a key issue for planners at both District and County level.
Southwest Water is responsible for the management of water resources over most of Devon and Cornwall, and this is examined with reference to present and future requirements. Bord na Mona is responsible for the management of all of the peat resources of the Irish Republic, and this is examined as a national case study, which also considers some of the conservation issues associated with Irish peat.

Managing Mineral Resources in the Isle of Purbeck

Figure 4.3.1 is an extract from a local newspaper in the Isle of Purbeck.

STUDENT **4.2** ACTIVITY

1 Identify the main points of concern that are raised in the article.
2 What are the main difficulties in achieving a careful balance 'between the disruption caused to local people, the damage done to the environment, and the local or national need for the mineral concerned'?

Figure 4.3.2 shows the main areas of mineral extraction in the Isle of Purbeck, and the adjacent Frome and Piddle valleys. It illustrates the remarkable concentration of working of different mineral resources within a relatively confined area. The working of Purbeck and Portland Stone in the south of the Isle of Purbeck, together with the extraction of ball clay in the heathland areas of the north have both been traditional well-established industries. One principal consequence of the longevity of these industries is that

'Don't turn Purbeck into an industrial wasteland,' pleads County Councillor

COUNTY COUNCILLOR Phil Duffy from Corfe Castle is urging Dorset County Council to radically rethink its attitude towards oil and other mineral exploration and extraction throughout Dorset – before, he says, the Isle of Purbeck is irretrievably despoiled and turned into a vast area of industrial wasteland.

Cllr Duffy says that he is horrified at the speed at which oil, clay and gravel is being extracted, which is causing great unease amongst local people.

He is urging that the county council adopt a co-ordinated policy and decide on a strict priority list for the extraction of minerals in sensitive areas so that a careful balance is achieved between the disruption caused to local people, the damage done to the environment, and the local or national need for the mineral concerned.

He told the *Purbeck Independent*: "Purbeck is sitting on a lot of valuable minerals in an areas of outstanding natural beauty, and if the extraction of oil, clay and gravel continues at its present rate then the area will be ruined. We have got to a stage now where there is a long list of applications for mineral extraction from different companies. There must be some structure and policy whereby they can be slotted into an agreed programme so that future generations will be able to enjoy our beautiful natural heritage as we do today."

Cllr Phil Duffy issued his warning after Dorset County Council's planning sub-committee decided to give BP the go-ahead to drill for oil in the hamlet of Organford Farm, just west of Lytchett Minster.

"We have seen what has been happening at Ridge with oil derricks being moved in and out of the area. We have seen the amount of inconvenience that the laying of oil and electrical pipelines from Wytch Farm to Furzebrook has caused local people," he said.

"We now see an operational gas loading terminal at Furzebrook where the residents have been issued with an emergency plan for evacuation if anything was to go wrong! It is all very well for other people in other parts of Dorset to say that everything is all right in Purbeck – and I don't deny that BP do their very best and have high standards of safety – but the march of progress that they are making throughout the Purbeck area is frightening. It is incredible how fast they are going.

"And it's not just BP. There are two applications from English China Clays; one for the extraction of ball clay at the Squirrel Cottage mine at East Holme that could involve the likely destruction of three ancient monuments and a valuable piece of heathland which is classified as an SSSI – and another for a huge gravel pit at Bestwall, the area of which is larger than Wareham's old town.

"I understand, of course, that each application must be judged on its own merits, but it must be within an agreed framework and overall policy for the future. Cllr Duffy ended: "DCC's Mineral Plan has been approved but it doesn't categorise any particular source or mineral and it doesn't put them in any order. It just states what must be done to protect the environment. That is not enough. We must act before it is too late."

Purbeck Independent 12 December 1990

Figure 4.3.1 *Don't turn Purbeck into an industrial wasteland*

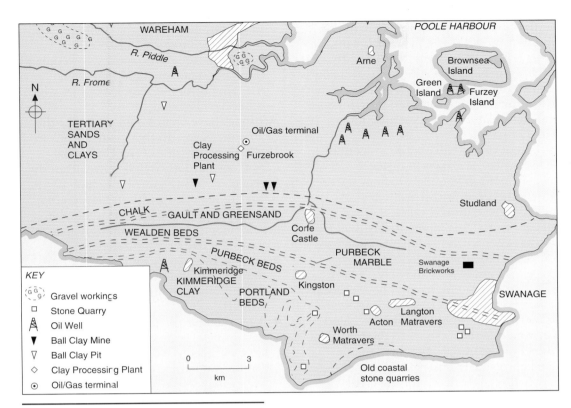

Figure 4.3.2   *Mineral exploitation, Isle of Purbeck*

controls in the past were far less strict than they are today. This has meant that parts of Purbeck, an Area of Outstanding Natural Beauty, are still scarred with derelict workings: this is particularly the case in parts of the heathland to the south and south west of Wareham, and on some parts of the limestone plateau to the west of Swanage. The demand for aggregates has increased steadily in the last two decades, and this has seen a parallel growth in the gravel extraction industry in the Frome and Piddle valleys. The discovery (1974

and 1978) and extraction of oil and natural gas (commenced 1979) in the Wytch Farm oilfield have added a further dimension to mineral activity in Purbeck.

BALL CLAY EXTRACTION IN THE PURBECK HEATHLANDS

Ball clay is a relatively rare mineral in Britain, and only occurs at three locations, two in Devon and one in the Isle of Purbeck in Dorset. Figure 4.3.3 shows ball clay production in Britain and indicates

	North Devon	South Devon	Dorset	Total
1983	140 206	339 414	100 651	580 271
1984	157 704	343 573	114 500	615 777
1985	143 319	346 174	97 528	587 021
1986	153 472	358 324	91 207	611 033
1987	172 105	385 124	112 289	669 518
1988	123 709	466 200	125 635	715 544
1989	131 413	511 278	136 009	778 700
1990	125 966	534 714	144 852	805 532
1991	–	–	–	728 575
1992	116 819	496 242	131 317	744 348
1993	126 876	483 490	135 465	745 831

Figure 4.3.3   *United Kingdom, sales of ball clay*

that the South Devon Basin is the most important producer (64 per cent), with the Dorset Basin (18 per cent) only slightly ahead of the North Devon Basin. Ball clay is a fine grained, plastic clay, and is widely used in the ceramics industry, for refractory manufacture and as an inert filler for a variety of products.

The extent of the ball clay deposits in the Isle of Purbeck is shown in Figure 4.3.4. The consultation boundary defines the area within which the British Ball Clay Producers Federation is notified of any applications for development. The ball clay deposits occur within sediments known as the Poole Formation, which consist of interbedded sands, silts and clays. The ceramics industry has demanding technical specifications for the clays that it uses, and this means that the industry has to blend clays from different sources to produce the required raw material. The most highly valued clays in the Purbeck area are those immediately to the north of the chalk ridge, in the Holme area, and in the vicinity of Trigon Hill to the north of the River Frome.

The Ball Clay Standing Conference in 1950 drew two main conclusions:

1 'ball clay was a mineral of national importance, of high export value and comparatively rare occurrence, not only in Britain but also in the world;

2 in all three producing districts the general nature of the surface is woodland or poorer quality agricultural land, implying little foreseeable conflict with other development.'

Attitudes have changed greatly since the Conference came to these conclusions in 1950. Sixty per cent of the Dorset Ball Clay Consultation Area is now within the Purbeck AONB (Figure 4.3.4) and 30 SSSIs are included within it also. Current strategy supports the idea of sustainable development in the Consultation area, and requires the ball clay producers to:

● meet as much as possible of the future needs of the industry from that part of the consultation area lying outside of the AONB;

● meet the unavoidable need for remaining clays from the least damaging sites within the AONB. Permission will not be granted for workings within, or where they would adversely affect SSSIs.

Current workings of ball clay are shown in Figure 4.3.4. It will be noted that all of the workings with the exception of Trigon Hill are within the AONB. In order to put into practice the above strategy for sustainable development, a **sieving exercise** was carried out to determine the least damaging options for the future. Three

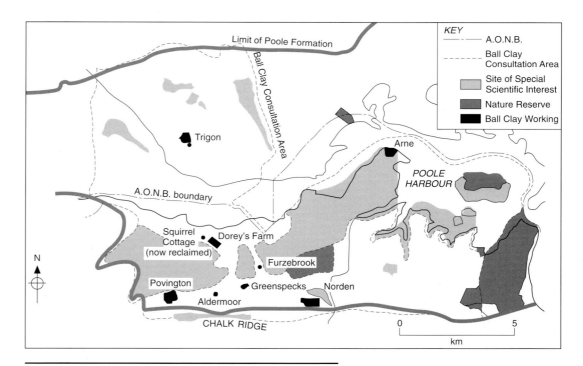

Figure 4.3.4 *Ball clay workings and environmental considerations*

sites were identified at Trigon Hill, Squirrel Cottage and Povington. All of the ball clay produced in the Isle of Purbeck is transported by lorry to the central processing plant at Furzebrook (Figure 4.3.5). It is a permanent industrial site within the ACNB, and its centralised function does prevent the proliferation of other similar sites in Purbeck.

STUDENT 4.3 ACTIVITY

1 Study Figure 4.3.4 showing the main locations of environmental value in the Ball Clay Consultation Area. Explain why Trigon, Squirrel Cottage and Povington were selected as the three areas for future development that were likely to be the least damaging.
2 What is likely to be the main environmental impact of the Furzebrook processing works? What measures could be taken in order to reduce its impact?

Figure 4.3.5 Ball clay processing plant, Furzebrook, Isle of Purbeck

Developing the Squirrel Cottage Site

Figure 4.3.6 is an article from the Purbeck Advertiser of October 1990, which describes the first proposals for the extension of the original Squirrel Cottage site.

Clash over clay works

PLANS to expand ECC's clay site south of Wareham have met with opposition from conservationists and archaeologists.

The company has applied to Dorset County Council for planning permission to extend its Squirrel Cottage site off Holme Lane, near Stoborough.

English China Clay has been working the rare clay from the 37-acre site since 1965. There is only one year's approved supply left, so the company wants to get the go ahead to excavate reserves in another 45 acres to the south of the site.

This would guarantee production for a further 10 to 12 years.

The plan entails:
* Re-locating part of the 20-acre Doreys Heath which is a Site of Special Scientific Interest (SSSI).
* Excavating three scheduled Bronze Age burial mounds.
* Felling a commercially-grown conifer woodland.

ECC says when it's finished restoring there will be three times more heathland, a new broadleaf woodland area, a large lake for wildlife and eight areas of pasture.

Philip Chesterfield, estates surveyor for ECC, told the Advertiser: "Animals graze on the heathland and they hold clay pigeon shoots on it.

"No smooth snakes or sand lizards (protected species) have been seen on the heath over the last 12 months."

He said: "The barrows have been dug into in the past and trees are growing on them. Our archaeological consultants say they are not of very high quality and they're deteriorating.

"If we don't get approval to excavate the barrows, 220,000 tonnes of clay will be sterilised which is 47 per cent of the total reserves.

"One of the plan's principal aims is protection of the environment and we've gone to great lengths to study what goes on around us."

The Nature Conservancy Council looks set to make an official objection. Senior officer Andrew Nicholson said: "We are not convinced that re-locating and replacing the heathland is an adequate alternative to leaving it as it is.

"It is by no means prime habitat but heathland is rare. It's very difficult to prove the absence of rare species.'

He added: "We are also concerned about the possible effects on the water supply to the protected bog."

British Herpetological Society officer Doug Mills claims he saw a sand lizard on Doreys Heath this week and he intends to fight for the threatened heathland.

English Heritage is talking to ECC about the three ancient barrows. Inspector Paul Gosling said: "We appreciate that the clay is a scarce and important mineral but the barrows are nationally important monuments.

"All monuments are deteriorating. They aren't ideally managed but we'd prefer it if they were left alone.

ECC is putting on a public exhibition of the plans at its North Street office in Wareham today (Thursday) between 2.30 pm and 7.30 pm, and tomorrow from 2.30 pm to 8.30 pm.

Comments and objections can be made in writing to the county planning officer, Richard Townley, at County Hall, Dorchester, Dorset.

Purbeck Advertiser 4 October 1990

Figure 4.3.6 Clash over clay works

1 Draw up a table to show the different attitudes and values that are represented in this article.
2 Show how these different attitudes and values are likely to result in conflict.

Figure 4.3.7   *Restoration of ball clay workings, Squirrel's Cottage, Isle of Purbeck*

This proposal was refused on the grounds that it could not be reconciled with the Dorset County Structure Plan. A new proposal was later submitted for a site to the south east of Squirrel's Cottage (now in the process of being restored, see Figure 4.3.7) where environmental damage was likely to be much less. Good quality heathland is not affected, no SSSI is involved, and the main archaeological site, Three Lord's Barrow is just outside of the workings. Human use of the area remains undisturbed. Planning permission was granted for this second application, and the new site is now one of the most important producers of ball clay in Purbeck (Figure 4.3.8).

Figure 4.3.8   *Dorey's Farm Ball Clay Pit, Isle of Purbeck, Dorset*

MANAGING THE WYTCH FARM OILFIELD

The Wytch Farm oilfield in the northern part of the Isle of Purbeck has seen a steady expansion since its first discovery in 1974. Oil was first discovered in the Bridport Reservoir at a depth of 900 m, and four years later a further discovery of oil was made in the Sherwood Reservoir at 1600 m. The extent of the two reservoirs is shown in Figure 4.3.9. In May 1984 BP became the operator of Wytch Farm Oilfield, and the main production features are shown in Figure 4.3.9. Most of the oil wells are located near the southern fringe of Poole Harbour (Figure 4.3.10) with two wells on Furzey Island, which BP purchased in 1983. Oil from the wells is piped to the central gathering station, where gas and water are separated from the oil. Oil is then piped from the gathering station to the terminal at Hamble on Southampton Water, with gas being delivered to the national gas grid in a separate pipeline. Liquified petroleum gas is sent by rail from the Furzebrook terminal.

Most recent developments in the oilfield have focused on the discovery of an extension of the Sherwood reservoir eastwards under the waters of Poole Bay in 1988. A range of options was considered for the recovery of this oil, and an artificial island, constructed in Poole Bay, was considered to be the preferred option. However there were significant environmental and economic disadvantages in the idea of an artificial island (although the method is used in a number of oilfields). Recent technological developments have persuaded BP that it would be possible to recover at least 80 per cent of the oil from a land-based drilling site, using the new technique of extended reach wells. Lateral drilling has extended as far as six kilometres in the North Sea, Australia, and the USA and BP considered that a laterally drilled well on the Goathorn Peninsula (Figure 4.3.11) could reach out to five km and possibly further.

Planning and Environmental Management of the new Well Site on the Goathorn Peninsula

The Goathorn site (see Figure 4.3.12) that was selected for the new lateral drilling complex is located in one of the most sensitive locations in Dorset. Inevitably, very strict planning regulations control

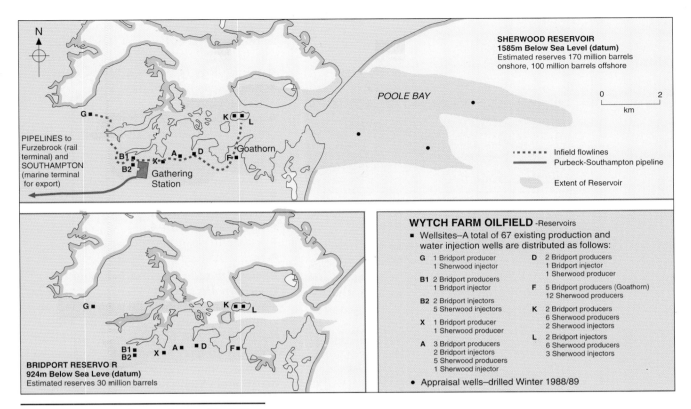

Figure 4.3.9   Oil and gas production, Wytch Farm

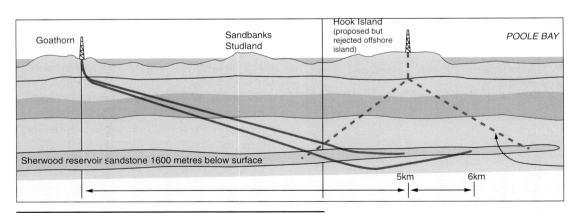

Figure 4.3.10   Oil Well, Wytch Farm, Dorset

mineral workings and development at such sites. Dorset County Council state that 'Within Areas of Outstanding Natural Beauty, Special Protection Areas, **Ramsar** sites, National Nature Reserves and Sites of Special Scientific Interest proposals for new mineral working or extensions to existing sites will be subject to the most rigorous examination.' Similar policies are expressed in the Local Plan of Purbeck District Council. In the production of the required Environmental Statement, due

Figure 4.3.11   Lateral drilling beneath Poole Bay from Goathorn

Figure 4.3.12 Goathorn Peninsula with drilling rig

consideration had to be given to a number of significant issues that were likely to arise in the light of the planning regulations and constraints imposed by the local Authority

Plans. The following environmental issues were identified as being of fundamental importance:

- the ecology of the Goathorn Peninsula
- landscape and visibility of the development
- noise incidence and management.

Other issues that were thought to be significant were:

- atmospheric quality
- hydrology
- archaeology
- neighbourhood and amenity
- traffic assessment
- incident assessment
- waste management.

Ecology, Landscape and Noise on the Goathorn Peninsula

The Goathorn Peninsula is protected by a number of statutory designations. The Peninsula is part of the Rempstone Heath SSSI which has characteristic bog and wet heath communities, with transitions to salt marsh and extensive mudflats in Poole Harbour, which is also a SSSI. The mosaic of habitats on the peninsula also includes dry heath, acidic grassland and broad-leaved woodland. Rare species that are found on the Goathorn Peninsula include the Dorset Heath, the sand lizard, the smooth snake and the nightjar. The peninsula fulfils the criteria for an EU Special Protection area. There is little agricultural activity on the peninsula, since it is limited to two rough pasture fields. Clearly there is much variation in the quality and diversity of the biological environment. Main ecological interest is in the rare and mobile reptile and bird species and the patches of various types of heath. The main features of the habitats are shown in Figure 4.3.13.

From a landscape point of view Goathorn is a wooded peninsula that projects into Poole Harbour. On the west it meets the Harbour in a series of low, but steep cliffs, and slopes gently down to the Harbour on the eastern side. The well site itself is well screened by mature coniferous woodland. Thus the essential characteristics of the peninsula of pine and deciduous woodland, heath, saltmarsh, mudflat and tidal water are unlikely to be spoilt by the construction of the well site. Visibility tests from a number of viewpoints indicate that the permanent facilities on the well site will not be seen.

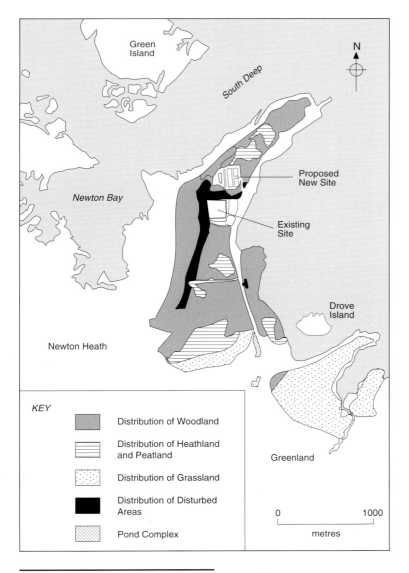

KEY

▨	Distribution of Woodland
▤	Distribution of Heathland and Peatland
▨	Distribution of Grassland
■	Distribution of Disturbed Areas
▨	Pond Complex

Figure 4.3.13 Goathorn, main habitats

In a quiet rural area it was essential that noise assessment was carried out. Prediction of noise levels was done by means of a computer model, and shown as a series of contours (Figure 4.3.14). No property came within the limiting 40 **dBA**.

Other Significant Environmental Issues
Tests for atmospheric quality have been carried out by BP for locations on the Goathorn Peninsula. Readings of sulphur dioxide, nitrogen dioxide and atmospheric hydrocarbon showed levels well below the calculated environmentally safe limit. **Lichen tests** gave similar results. Construction, drilling and operation of the well site should have no adverse effects on the hydrology of Goathorn Peninsula. Concern has been expressed over the possible contamination of groundwater, since tarmac would reduce the percolation of rainwater, and might lead to the invasion of the freshwater below the water table by saltwater from Poole Harbour. This has not happened at the well site on the neighbouring Furzey Island and is unlikely on Goathorn. Unlike some of the other peninsulas on the southern shores of Poole Harbour, Goathorn appears to have little in the way of archaeological sites. Goathorn has no public rights of way, and construction and operation of the well site are thus unlikely to cause little more than modest disturbance to people in the area. Amenity interests within Poole Harbour are protected by screening.

Access to the Gathering Station is by private road from the main A351 into the Isle of Purbeck from Wareham. From the Gathering Station unmade tracks lead to the well sites including the Goathorn site. As construction involves the moving of much heavy equipment along this track, it has been strengthened with stone, due regard being paid to the sensitive nature of the heath on either side.

Pollution contingency planning is undertaken by BP for all of its sites, but is particularly important at Goathorn owing to its nearness to Poole Harbour. A contingency plan exists already for the existing well site on Goathorn, and the new site will share some of the features of the existing one. Waste management will be dealt with through existing procedures: the main problem will be the disposal of drilling mud and this will take place through one of the **injection wells** on the field.

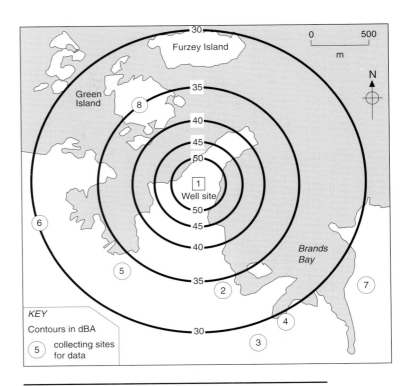

Figure 4.3.14   *Noise contours for the new well site at Goathorn*

STUDENT   **4.5**   ACTIVITY

1 Examine each of the issues identified by BP as being important at Goathorn. For each issue
   a state why it is important as a matter of environmental concern
   b discuss what reassurances the oil company would have to give to the local planning authority in order to convince them of the environmental safeguards they have put in place.

AGGREGATE DEMAND: THE CASE OF SWINEHAM FARM

Aggregates are angular fragments of inert materials that are usually bound together by cement or bitumen to produce concrete or tar macadam. Three main sources of aggregate exist in Dorset: land-won sand and gravel, marine dredged sand and gravel and crushed limestone rock (Figure 4.3.15). It will be noted that there is an important concentration of these materials in the Isle of Purbeck. The main gravel bearing formations are the Plateau Gravels and the Valley Gravels, which, in the

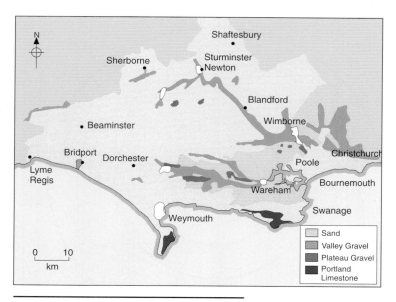

Figure 4.3.15   *Aggregates distribution in Dorset*

Figure 4.3.18   *Gravel workings at Swineham Farm, Wareham, Dorset*

Wareham area, were laid down by the Rivers Frome and Piddle.

Figure 4.3.16 shows aggregate production in Dorset over the period from 1980–91 (the drop in production resulting from falling demand due to recession can be

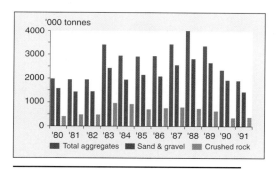

Figure 4.3.16   *Aggregates production in Dorset*

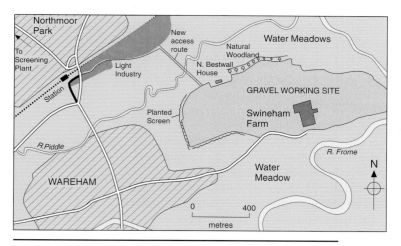

Figure 4.3.17   *Gravel production site at Swineham Farm, Wareham*

seen at the beginning of both the 1980s and 1990s). In the period 1992–2008 there is a demand level for 11.2 million tonnes of gravel, but existing permitted reserves are only of the order of some 3.5 million tonnes, leaving a shortfall of 7.7 million tonnes. This shortfall means that new planning permissions are required to make good the balance.

Gravel at Swineham Farm

The Swineham Farm site for gravel extraction is shown in Figures 4.3.17 and 4.3.18. The site occupies some 55 hectares, and the proposal envisaged the extraction of some 1.6 million tonnes of gravel over a period of ten years. The site is relatively close to the old walled town of Wareham, and access to it was only available through Wareham. The developers thus proposed a new access route through the wetlands of the River Piddle to the north (Figure 4.3.19), which would give a direct link between the pit and the screening plant at Tatchells to the north west of Wareham. This route would cross a busy main road and pass through a large residential area, with an estimated 50 lorry loads of gravel being transported each day.

There was considerable public outcry over the proposal, and Dorset County Council, Purbeck District Council and Wareham Borough Council all opposed the application, and it was refused planning permission. However the developers appealed, and their appeal was upheld by the Secretary of State for the Environment. Production began in 1992, with a pit life of ten years.

Working of this pit involves the removal of overburden, and its stockpiling before restoration takes place. The developers are obliged to return the worked area to

Figure 4.3.19   *Access road to Swineham Farm Gravel Workings*

agricultural use, although the extraction of up to five metres of gravel means that the restored height is some two metres below the original. Final restoration plans are shown in Figure 4.3.20. The lakes will offer new habitats for a wide range of flora and fauna.

Management of the working involves a number of different aspects:

- supervision of the removal and stockpiling of the overburden;
- monitoring of the 100 lorry movements per day (suspended between 8 am and 9 am on schooldays);
- monitoring of noise levels on the site;
- monitoring of the water table and water quality;
- supervision of restoration work, which includes both replacement of the

overburden, and the reseeding of the topsoil.

Apart from the obvious economic benefits of gravel extraction, two other positive aspects of the operation are apparent:

- important archaeological finds have resulted from the stripping of the overburden, and the archaeologists are given unlimited access to the site for research;
- the enhancement of wild life in the area, with the final creation of large water bodies as a part of the restoration programme.

STUDENT 4.6 ACTIVITY

1 This section has examined three case studies of mineral extraction in the Isle of Purbeck

   a Return to Student Activity 4.2. How successful do you think the local minerals planning authority has been in achieving 'a careful balance between the disruption caused to local people, the damage done to the environment, and the local or national need for the mineral concerned'?

   b Attempt to rank the three case studies in the way that they have achieved the balance between the different elements referred to above. An environmental impact matrix could be used to assist the ranking.

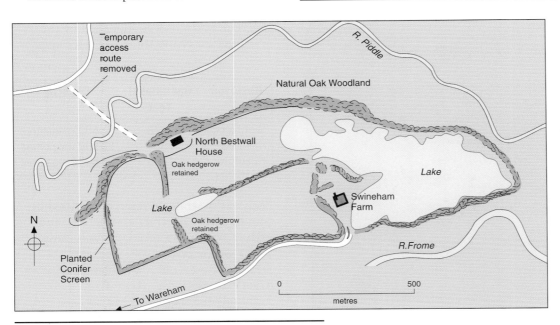

Figure 4.3.20   *Restoration Plan, Swineham Farm gravel working*

4.4 Managing Water Resources

WATER IN THE SOUTH WEST

South West Water serves an area of nearly 11 000 km² in Cornwall, Devon and small areas of Dorset and Somerset, and serves a resident population of over 1.5 million people, with an additional peak visitor population in the summer of 500 000 people. Water to supply these consumers comes from three sources: rivers, reservoirs (see Figure 4.4.1) and groundwater supplies (see Figure 4.4.2). Ultimately all three of these sources depend on replenishment from rainfall, and the south west of England generally receives adequate rainfall for this purpose. Figure 4.4.3 shows a series of rainfall graphs for a variety of locations in the south west.

Figure 4.4.1   Roadford Reservoir, Devon

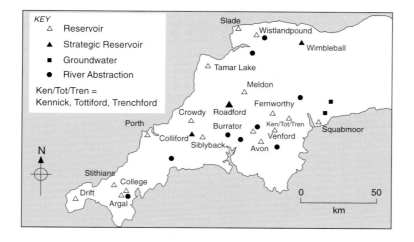

Figure 4.4.2   Sources of water supply, South West Water

STUDENT 4.7 ACTIVITY

1 Comment on the spatial variations of rainfall over the south west of England.
2 Calculate the mean annual rainfall figure for the 12 stations shown on the map, and the standard deviation. What does this tell us about rainfall variation in the South West.
3 What implications do the results of the above calculations have for water supply in the South West?

THE WATER SUPPLY NETWORK

Figure 4.4.4 shows the main features of the water distribution network in the South West. It will be seen from the map that there are six main impounding reservoirs in the region. The 'big three', Wimbleball, Roadford and Colliford act as the strategic water store of the south west. When South West Water first took over responsibility for water resources in the South West in 1974, there was a major water storage problem, and this was addressed by the construction of the three key reservoirs.

Reservoir	Operational Year	Capacity
Wimbleball	1978	21 500 megalitres
Colliford	1984	28 500 megalitres
Roadford	1990	34 500 megalitres

Each of the strategic reservoirs is used in conjunction with river abstraction, and with other smaller reservoirs as follows:

The Exmoor system
Strategic Reservoir: Wimbleball
River Abstractions: River Exe
Total reliable supply: 60 Ml/day
Area served: Exeter and towns within Exe and Culm catchments, but also acts as back

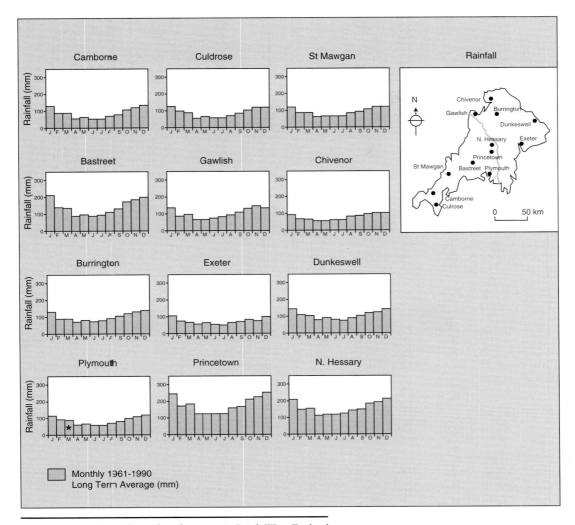

Figure 4.4.3 *Rainfall at selected stations in South West England*

up to groundwater supplies in the Otter and Axe catchments in the east of Devon. Also supplements North Devon with transfers possible to the Taw catchment

The Cornwall Reservoirs
Strategic Reservoir: Colliford
River Abstractions: Rivers Fowey, Camel and Lynher
Local Reservoirs: Argal and College, Crowdy, Drift, Stithians and Porth
Total reliable supply: 105 Ml/day
Areas served: Cornwall

The Dartmoor Reservoirs
Strategic Reservoir: Roadford, in conjunction with Meldon and Burrator
River Abstractions: Tamar, Taw, Dart and Tavy
Local Reservoirs: Avon, Fernworthy Kennick, Tottiford and Trenchford
Total reliable supply: 255 Ml/day

Area served: North Devon, South Devon and Plymouth

Capacity and percentage full figures for the South West for mid March 1996 are shown in Figure 4.4.5.

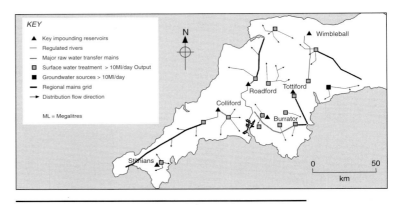

Figure 4.4.4 *Water distribution network, South West Water*

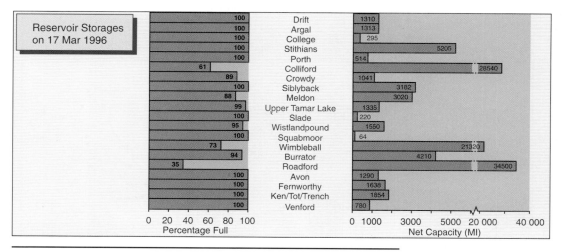

Figure 4.4.5 *Capacity and percentage full figures for South West Water reservoirs*

STUDENT **4.8** ACTIVITY

1 Comment on the hierarchy of reservoirs in the south west as shown in Figure 4.4.2.
2 Comment on the supply situation as shown in Figure 4.4.5, remembering that 1995 was a particularly dry summer.

Assessing Future Demand

Figure 4.4.6 shows future levels of demand against current resources in South West Water's region. The table (Figure 4.4.7) shows the predicted annual Public Water surplus and deficits for the South West Water Region for 2001, 2011 and 2021.

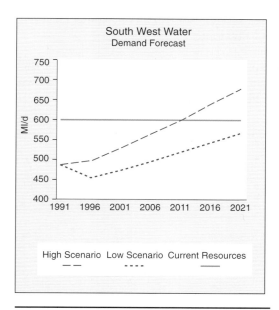

Figure 4.4.6 *Future levels of demand against current resources for South West Water*

STUDENT **4.9** ACTIVITY

1 Comment on the future surplus/deficit patterns for the three strategic reservoir regions as shown in the diagram.

Where present resources appear unlikely to be able to support future demand, three options are available (in order of decreasing preference)

● demand management
● resource management
● resource development.

Demand management: two main options exist to control demand
1. water metering: in the whole of the South West Region of the Environmental Agency, (which also includes Somerset, Avon and parts of Wiltshire and Hampshire) all commercial and industrial consumers pay for their water supply through metering, less than 3 per cent of domestic properties are currently metered. It is likely that metering could make considerable savings. A national trial reveals up to 7 per cent savings, and in the Isle of Wight figures as high as 22 per cent have been suggested. South West Water currently meters all new businesses, but this does not include all new domestic properties. The Environmental Agency concludes that metering may be economic in areas where:

● peak demands are particularly high due to seasonal factors;
● new sources cannot be developed in time to meet the onset of a resource deficit;

Strategic Reservoir Zone	2001		2011		2021	
	High	Low	High	Low	High	Low
South West Water Colliford Zone	−3	18	−29	2	−57	−17
South West Water Roadford Zone	54	83	18	61	−21	35
South West Water Wimbleball Zone	2	6	−10	−2	−22	−10

Figure 4.4.7   *Predicted annual public water supply surpluses and deficits in South West Water's strategic reservoir zones (Ml/d)*

- new sources can only be secured at high capital or environmental cost.

2. efficient water use: there appear to be three main ways in which this might be achieved:

- voluntary restraint in the use of water achieved through better public information;
- the development and public acceptance of domestic appliances and plumbing systems that are more efficient in their water use;
- the promotion of more efficient industrial use of water, e.g. recycling and re-use.

Resource Management: six main options exist here:

- leakage control – water lost from overflows, burst pipes, leaking joints and dripping taps South West Water has programmes in hand to reduce leaks from its pipes to 15 per cent by 2000 compared to the current level of 24 per cent;
- operational improvements: these include conjunctive use of resources, e.g. pumping water from the Exe into Wimbleball reservoir in the winter to ensure that the reservoir has the capacity to augment flow in the Exe in the following summer;
- resource integration: interlinks between existing resources should enable areas with surpluses and areas with deficits to be more effectively independent. In the South West additional links between the three strategic areas would generally improve efficiency;
- inter-company transfers: surplus and deficit areas that exist between water companies could be equalised by inter-company transfers;
- effluent re-use: suitably treated waste

water can be returned to rivers so that it can be abstracted downstream;
- abstraction licensing procedure: new abstraction licences require that the licensee makes a statement on the effect of the abstraction on the environment. All new licences are now subjected to very rigorous environmental scrutiny, and the effect of resources is closely monitored.

Resource Development
Demand and resource management should, in the South West Water region restrict demand to the low level, i.e. 10 Ml/day for Wimbleball and 17 Ml/day for Colliford. It would appear that the best options for the three strategic reservoirs in the South West would be some form of pumped storage scheme, such as that described above for the Wimbleball Reservoir. In this way both the deficiencies in the Wimbleball and Colliford strategic regions could be met.

STUDENT   **4.10**   ACTIVITY

1 Review the forecast water situation in the South West Water region.
2 Examine all of the options for reducing the potential deficits in the twenty first century, and suggest what would be the best option for making savings.
3 Read the newspaper extract from the *Irish Times* 'Paying for Water' (Figure 4.4.8).
   a What are the main issues involved in the supply of water to the Dublin area?
   b What are the main solutions proposed to current problems?
   c To what extent do they reflect similar problems in England and Wales?

Paying For Water

THE publication this week of a comprehensive study on Dublin's water needs could hardly have been more timely, after the recent Dublin West by-election pushed the water-charges issue up the political agenda. The report is unequivocal in its support for much more widespread water charges and signals that the installation of water meters in each home may be the only way to finance the water system into the next century.

On the basis of the report, few could deny the need for major investment in water projects. The most startling fact to emerge is that about 160 million litres of water per day – almost half of the Dublin area's total supply – is literally poured down the drain because of widespread leaks in the distribution network. This unacceptably-high level of leakage is significantly more than in any other European capital and provides damning evidence that the system is overwhelmed by the demands made upon it.

The task now is twofold: investment is required to plug the leaks and to expand the supply network. But the cost is also prohibitively high: the consultants estimate that over £500 million will have to be invested in major water projects in the Dublin area over the next twenty years, of which £160 million will be required by the year 2000.

* * *

The question, as always, is who should pay for this investment? Clearly the present system, in which major cities like Dublin and Limerick are exempt from water charges while other areas pay anything between £25 and £150 per year is unsustainable in the longer term. It makes no sense to ask one sector of the community to pay for a service which others can enjoy without charge. There is also the conservation issue: a free water supply like that in Dublin city does little to encourage water conservation. And there is the equity issue: it hardly seems fair that every household in any one council or corporation area pays the same water charge irrespective of use.

The case for uniform water charges in the short-term and a move to water meters in each household in the longer-term, is persuasive: this is perhaps the best and most equitable means of paying for the investment that the water system so badly needs. Those who have campaigned so vigorously against the water charges would see any such regime as an additional form of taxation. They are right when they point to the grossly inequitable taxation system in this State, to the undue burden which falls on the PAYE sector and to the manner in which many of those best able to pay, appear able to minimise their tax liability.

They have a well-grounded case for tax reform but this should be made on its own merits: it should not be used to delay the introduction of a sensible and much-needed service charge. Whether the politicians have the courage to address the issue is another question. After the publication of this week's report, the Minister for the Environment, Mr Howlin, said pointedly that "politicians will reflect on the views expressed by the voters in Dublin West". It is to be hoped that these comments are not a prelude to more fudge and compromise. The water issue needs to be fully addressed – without any further delay.

Irish Times 12 April 1996

Figure 4.4.8   *Dublin: Paying for water*

4.5   Managing Peat Resources in the Irish Republic

Peatland occupies 16.2 per cent of the island of Ireland, and 17.2 per cent of the area of the Republic (see Figure 4.5.1). Only three countries in the world have a greater covering of peat (see Figure 4.5.2). In an island conspicuously lacking in other energy resources, the use of peat as a fuel in energy production has achieved an unusual prominence. In addition to its energy role peat is produced for a wide range of other purposes in the Republic of Ireland. Its national importance is such that peat resources are managed by a state company,

Bord na Mona (formed in 1946), which is responsible for the overall development of Irish peatlands as a commercial resource. However, the ecological richness of Ireland's peatlands could be put at risk in future developments, and conservation has now become a major issue in these areas. In 1982, the Irish Peatland Conservation Council was founded to promote the conservation of Irish peatlands through a programme of lobbying, publicity, site purchase, and educational programmes.

THE DISTRIBUTION AND ORIGIN OF IRISH PEATLANDS

Three main types of peatland are found in Ireland (see Figure 4.5.1).

- fens: undamaged fenlands are relatively rare in Ireland today, largely as a result of land reclamation schemes. They are found in only a few locations in the Irish Midlands, but do also occur in lowland areas in the west of Ireland. Black peat found in the fens is formed from the remains of sedges and fen mosses which grow in alkaline conditions.

- raised bogs (see Figure 4.5.3): these are found mainly in the Midlands of Ireland, in areas that have 800–900 mm of rainfall. They develop from fens and have a domed shape surface, and consist mainly of moss peat, formed principally from Sphagnum moss. These are some of the deepest of the Irish peatlands, with depths of over 13 m recorded in some locations. Raised bog development began at the end of the last glaciation in Ireland, about 10 000 years ago. The main stages of development are shown in Figure 4.5.4.

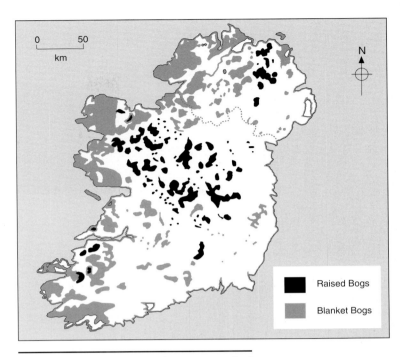

Figure 4.5.1 *Distribution of bog types in Ireland*

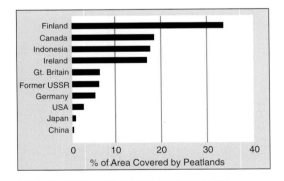

Figure 4.5.2 *Worldwide peatland resources*

Figure 4.5.3 *Raised bog being worked for peat, County Offaly*

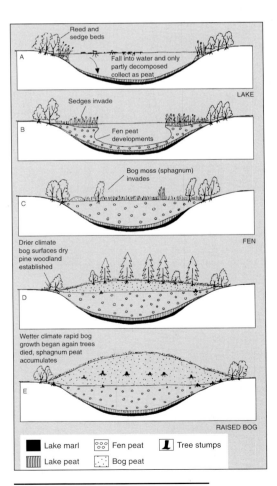

Figure 4.5.4 *Development of raised bog*

Figure 4.5.5   Blanket bog, Connemara

- blanket bogs: much of the western coastlands of Ireland carry a cover of peat varying from two to six metres in depth (see photograph Figure 4.5.5). This is known as Atlantic or Low level blanket bog and is formed largely of the remains of grasses and sedges, and contains very little bog moss. It develops in areas where there is an excess of 1200 mm of rainfall. High or Mountain blanket bog is found on flat or gently sloping areas in mountains throughout Ireland, usually above 200 m. Blanket

bog formation also began at the end of the last glaciation in Ireland. Initially, peat formation was confined to shallow lakes and hollows, but later acid peat spread out over huge areas, particularly after 4000 years ago when the climate became wetter.

THE EXPLOITATION OF PEAT AS A RESOURCE

Peat has been used as a traditional fuel in rural Ireland for nearly a thousand years. Stacks of hand-cut sods of peat are still a characteristic feature of much of the remoter western seaboard of Ireland. However, hand-cutting of peat is now being replaced by mechanical means. 'Difco' peat cutting machines can now be attached to the back of tractors, rendering the production of peat for home consumption a far less labour-intensive operation.

Peat production first became commercially organised under the Turf Development Board in 1934, and this was replaced by the state organisation Bord na Mona in 1946. The main aims in setting up Bord na Mona were to develop and manage the

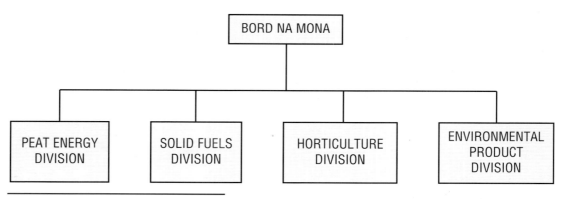

Figure 4.5.6   Organisation of Bord na Mona

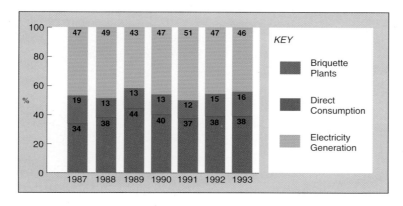

Figure 4.5.7   Consumption of peat by usage

Irish peatlands industry, and to produce and market peat and peat products. The organisation of Bord na Mona is shown in Figure 4.5.6. To date 88 000 hectares of bogland have been acquired by Bord na Mona and used for the production of peat for Irish electricity generation, a range of peat and peat-based fuels for domestic and industrial heating, horticultural peat for both home gardeners and commercial horticulturalists, and recently, pollution abatement materials.

The diagrams in Figure 4.5.7 show the

	Primary Generation (million units)	%
Hydro	723	5
Peat	2100	14
Coal	5831	40
Gas	3641	25
Oil	2410	16
Total Primary Generation	14 705	100
Pumped Storage (Generation)	218	
Pumped Storage (Pumping)	(311)	
Purchased	22	
Total System Demand	14 634	

Figure 4.5.8 *Sources of energy, Irish Republic*

STUDENT **4.11** ACTIVITY

1 Review the relative importance of the different categories of peat consumption.
2 Study Figure 4.5.8. Comment on the role of peat as an energy resource in the Irish Republic.

Bord na Mona currently supplies peat to five power stations in the Irish Republic (Figure 4.5.9). Their location, and the principal areas of operation of the company are shown in Figure 4.5.10. The first peat fired power station was opened at Portarlington in 1950. Although uneconomic at first, the social benefits, particularly the employment provided by Bord na Mona outweighed the economic disadvantages. Oil price rises in 1973–4 did however result in peat-fired power stations becoming economic. Peat, like any other fossil fuel, is finite, and it is thought that Bord na Mona production can continue for another 30 or so years. Much of the existing capacity in peat-fired stations is

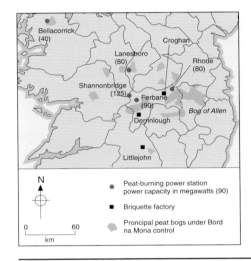

Figure 4.5.10 *Peat fired power stations and briquette plants*

now over 20 years old, and it would appear unlikely that new stations will be built in the future. As the older stations require major overhaul, it is likely that they will be decommissioned. It is perhaps significant that Bord na Mona has been involved in the earliest trials of wind-derived power at Bellacorick in County Mayo. Three small power stations in the west of Ireland are supplied from private sources of peat, hand won in some cases, but increasingly cut by machine.

Apart from the production of milled peat for the electricity generation industry, Bord na Mona has three other important operational divisions. Although the solid fuel market in Ireland appears to be in a general state of decline, briquette sales appear to be holding up well, although rationalisation of production has led to the closure of one plant recently. Horticultural

Figure 4.5.9 *Peat fired power station, Rhode, County Offaly*

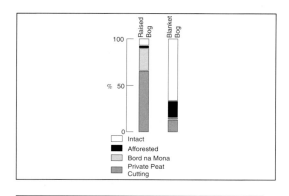

Figure 4.5.11 *Original and present extent of raised bogs in the Midlands of Ireland*

products are particularly important for their export sales, both within the United Kingdom and in continental Europe, although Ireland's peripheral position does put it at something of a disadvantage in this field. Environmental products are likely to be an important growth area in the future, with **biofiltration** of liquid effluents and gaseous emissions being particularly important.

An increasingly important issue for Bord na Mona is the use of the **cutaway bog** after peat production has ceased. Environmental policy now insists upon careful planning of the after use of cutaway bog. Testing of possible uses began as long ago as 1955. Several factors, such as the type of subsoil, the terrain beneath the bog and the drainage pattern of the cutaway will determine the future use. Landscape

and social considerations will also have to be taken into account, and current thinking envisages future land use as a combination of grass, forestry and wetlands, some of which will be for amenity use.

THE CONSERVATION ISSUE

Although the peatlands of the Irish Republic have been exploited for a very long period, the rate of loss of ecologically valuable land has accelerated considerably over the last ten years. In the case of raised bogs they originally covered 311 000 hectares in the Irish Republic, but by 1974 only 65 000 hectares of the original area were still relatively undisturbed or intact. However, a further survey in 1985 showed that the intact area had been reduced to only 20 000 hectares – about 7 per cent of the original area. In the Irish Midlands, where most of the commercial peat cutting takes place, the amount of intact raised bog was reduced from 39 000 hectares to a mere 11 000 hectares with nature reserve potential. County Louth and County Meath have no peatlands left. The map (Figure 4.5.12) shows the extent of the exploitation of raised bog in part of the Irish Midlands.

It would appear that private cutting has contributed most to the loss of raised bog, since it has been practised for such a long period. Bord na Mona has cut away nearly a quarter of the raised bog, but most of this loss has occurred in the last forty years. The Turf Development Act of 1981 made it possible for private individuals to obtain grants to extract peat. With the advent of the 'Difco' machines, this has meant that private peat cutting has affected almost every bog to some degree.

Blanket bog covered 772 000 hectares of the Republic of Ireland, and is now under threat not only from peatcutting, which has affected 15 per cent of its extent, but also from afforestation, which has affected a further 18 per cent. Much of the remaining area has been affected by over grazing and repeated burning.

Figure 4.5.11 shows the contribution of each activity to the loss of peatland in both raised and blanket bogs. Raised bog is currently being removed at a rate of some 3000 hectares a year. At this rate all of the potential nature reserves are likely to have disappeared by 1997.

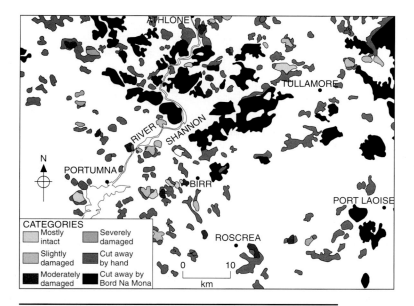

Figure 4.5.12 *Contribution of various activities to loss of peat bogs*

Legislation to protect the potential nature reserves within the Irish peatlands is embodied within the Wildlife Act of 1976. The Forest and Wildlife Service are responsible for the implementation of the Act. This body has the responsibility of identifying peatland sites with potential nature reserve status. The various locations are ranked according to their status – as being of international, national, regional or local importance. On its own, recognition of these sites does not afford protection, unless they are state owned. Funds for conservation have generally been in short supply, and have only been made available when there has been considerable public pressure for the protection of a particular site, such as Clara Bog in County Offaly. It is possible for management agreements to be arranged with private landowners, so that land use practices in areas surrounding peatlands, which might be detrimental to the peatlands themselves, might be avoided. However shortage of funding has generally made this an unlikely option in many cases. Indirectly, the Forest and Wildlife Service can ensure that Government or European Union funds are not made available for development of peatlands. This can be done through careful screening of grant applications to check for possible conflicts with peatland conservation.

Bord na Mona is now committed to protecting areas of peatland of conservation value. Working closely with the Forest and Wildlife Service, it is currently transferring large areas of peatland to public ownership. Furthermore it no longer acquires peatlands that are of agreed conservation value.

Apart from official Government bodies, there are a number of non Government organisations which have an interest in peatland conservation. The most important of these is the Irish Peatland Conservation Council, which works closely with a similar organisation in the Netherlands, the Dutch Foundation for the Conservation of Irish Bogs. The Conservation organisation known as An Taisce also has a committed interest in peatlands, and has raised funds to buy important conservation sites, such as the fen at Bellacorick, in County Mayo, close to one of the five peat-fired power stations.

STUDENT ACTIVITY

1 Assess the arguments for and against the continued exploitation of the valued national peat resource in the Irish Republic.

ESSAYS

1 To what extent is it true that environmental considerations are now foremost in determining whether a particular resource should be exploited?

2 National interest may often have to take precedence over local concern in resource development. To what extent is this true?

FIELDWORK OPPORTUNITIES & PROJECT SUGGESTIONS

Select a local quarry or pit where mineral resources are extracted. Establish contact with the manager, and discover if any relevant documents are available for you to consult: these could include the environmental statement submitted when the plan was submitted to the local authority for approval, or a site management plan.

Draw your own large scale plan of the quarry or pit to show the methods used for extraction and restoration. Investigate all of the different environmental impacts that the pit or quarry is likely to have. You will need to consider such things as visual appearance (you could sketch or photograph the location from different angles), noise pollution, water pollution, dust pollution, lorry traffic levels and so on. Management measures to lessen all of these impacts will then have to be investigated, and an assessment of their effectiveness made.

5 Managing Forests and Woodlands

Most woodland in this country has been managed or influenced by human activities in the past, and it still needs to be managed to maintain or develop the characteristics we value, such as the contribution to landscape, the wealth of wildlife it supports and the recreational enjoyment it provides.

Andrew Lane and Joyce Tait: Practical Conservation Woodlands

KEY IDEAS

- Woodlands and grasslands are both valued ecosystems in Britain and Ireland
- The present structure of these ecosystems has developed over a long period of time, and reflects both natural change and human influence
- The ecosystems have an important role to play as resources
- Management of woodlands and grasslands is concerned with maintaining a healthy ecosystem, capable of supporting sustainable development

5.1 Introduction

By international standards, both the United Kingdom (10 per cent) and Ireland (8 per cent) are not well endowed with forest and woodland, when compared to European Union countries as a whole (25 per cent), or the Scandinavian countries, Norway (27 per cent), Sweden (68 per cent) and Finland (76 per cent).

Sixty two per cent of forests in Great Britain are in the private sector, and this is a figure that is likely to increase. Eighteen counties and regions in Britain have seen more than a half of their public forests privatised since 1981 (see Figures 5.1.1 and 5.1.2). New planting in the private sector has far exceeded that in the public sector, as is evident in Figure 5.1.4. Much of this planting has been funded under the Farm Woodland Scheme, and the Farm Woodland Premium Scheme.

In both Northern Ireland, and in the Republic of Ireland forestry has predominantly been managed within the public sector for much of this century. Eighty per cent of the forest estate is in

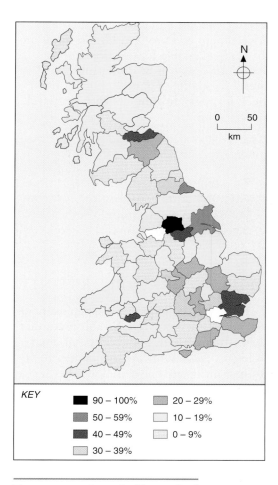

KEY

■ 90 – 100%	▨ 20 – 29%
■ 50 – 59%	▢ 10 – 19%
■ 40 – 49%	□ 0 – 9%
▢ 30 – 39%	

Figure 5.1.1 *Loss of public woodland*

	Total area (hectares) 1981	Number of woods 1981	Area (hectares) sold 1981–1996	Woods sold 1981–1996	% of area sold	% of woods sold
England	309 233	2456	33 550	999	11	41
Scotland	728 753	1704	75 501	820	10	48
Wales	149 725	1718	15 867	861	11	50
Total	1 207 711	5878	124 918	2680	10	46

Figure 5.1.2   *Summary of sales of woods by county*

public ownership in the Republic, with management of commercial aspects of the estate in the hands of Coillte Teoranta (the Irish Forestry Board). Promotion of private sector forestry is in the hands of the Forestry Service of the Department of Agriculture, Food and Forestry. Important changes are beginning to take place as a result of membership of the EU. In the Republic prices of farmland rose dramatically in the 1970s and this tended to discourage the purchase of land for public sector forestry. On the other hand European funding was available in the form of grant support for afforestation in the Republic of Ireland under the Programmes for Western Development. The aim of this scheme was to promote private forestry on land that is marginal for agriculture, but suitable for forestry. The level of grants is particularly generous, covering up to 85 per cent of the costs in the case of farmers and 70 per cent of the costs in the case of non-farmers. In Northern Ireland only 5 per cent of the land is under woodland (see Figure 5.1.5) and is currently managed by the Forest Service of the Department of Agriculture. There are plans to extend this forest cover by 120 000 hectares by the end of the century, with most of the extra land coming from marginal upland peat areas, where only a limited range of tree species can grow because of the harsh physical conditions.

Figure 5.1.3   *Private woodland: Lewesdon Hill, West Dorset*

Year ending 31st March	England		Wales		Scotland		Great Britain		
	FC	PW	FC	PW	FC	PW	FC	PW	TOTAL
1987	0.1	1.3	0.1	0.8	5.1	17.3	5.3	19.4	24.7
1988	0.2	1.8	0.2	1.0	4.6	21.2	5.0	24.0	29.0
1989	0.1	1.9	0.1	1.0	3.9	22.5	4.1	25.4	29.5
1990	0.2	3.5	0.1	0.6	3.8	11.5	4.1	15.6	19.7
1991	0.0	4.4	0.0	0.5	3.5	10.6	3.5	15.5	19.0
1992	0.1	4.1	0.1	0.4	2.8	9.8	3.0	14.3	17.3
1993	0.0	5.3	0.0	0.4	2.3	9.8	2.3	15.5	17.8
1994	0.1	6.1	0.0	0.6	1.3	9.2	1.4	15.9	17.3

Thousands of hectares

Notes:
1 The private woodlands figures are based upon the areas for which grants were paid during the year. They include grant-aided planting in association with the Farm Woodland Scheme, together with an estimate of areas planted without grants.
2 FC (Forestry Commission). PW (Private Woodlands).

Figure 5.1.4   *New planting, Forestry Commission and Private sector*

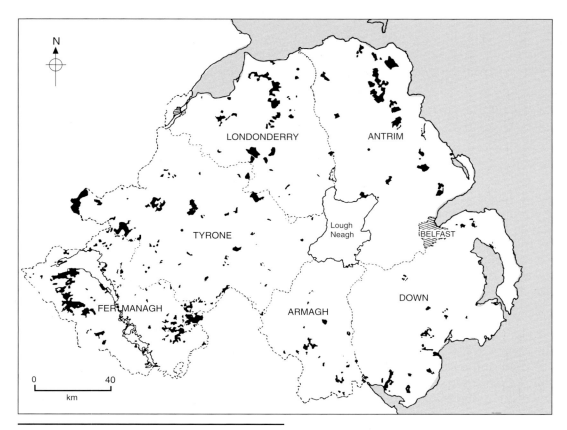

Figure 5.1.5   *Distribution of forests in Northern Ireland*

It is pertinent to review the role of the Forestry Commission in the management of the forest resources of Britain. The creation of the Forestry Commission in 1919 stemmed from concern over strategic timber supplies during World War I. Inherent in its establishment was the dual role it plays in management.

This dual function was more fully recognised in 1992, by the creation of two distinct bodies. The Forest Authority is responsible for the general regulation of and maintenance of standards in forest activity, and it also supervises grant provision, in particular the Woodland Grant Scheme. Forest Enterprise is responsible for the management of the Commission's forests and woodlands (see photograph, Figure 5.1.6), with the three principal responsibilities of timber production, the provision of recreational facilities, and conservation of the woodland environment.

Figure 5.1.6   *Coniferous plantations, Bennan Forest, Southern Uplands, Scotland*

Case Studies of Management of Forest and Grassland Ecosystems

5.2

It is possible to make the distinction between forest management in lowland Britain, where the emphasis is generally, but not entirely, on broad-leaved deciduous trees, and the management of upland conifer plantations in the higher parts of Britain. In England there are some 450 000 hectares of broadleaved forest, compared to 382 000 hectares of coniferous woodland. Reflecting the role of upland conifer plantations, Scotland possesses 966 000 hectares of conifers, compared to a mere 100 000 hectares of broadleaved trees. The New Forest has been chosen as a case study of a lowland forest area, since it represents a long established ecosystem in a variety of localised environments, forming a mosaic of woodland, grassland, and open heath. Its very longevity has led to the evolution of a complex web of management practices, that are in some ways unique to the Forest. More recently controversy over its possible designation as a National Park has finally resulted in 1992 in its being declared the New Forest Heritage area, with special status similar to a National Park.

By way of contrast, the Galloway Forest Park in the west of the Southern Uplands of Scotland encapsulates much of the character of upland forestry management.

Although some of the higher land in the north of the Park around the Merrick, is true open mountain wilderness, extensive coniferous plantations occupy much of the area. However, in sheltered areas, such as Glentrool, remnants of broad-leaved forest remain, and require their own distinctive style of management. Although it does not come under the same recreational pressures as the New Forest, with its adjacent urban areas, Galloway nevertheless has an important role to play in leisure provision in south west Scotland.

Figure 5.2.1 *Coniferous and deciduous woodlands: Loch Trool, Galloway Forest Park*

The New Forest: Management of a Diverse Ecology

5.3

The New Forest represents an assemblage of ecosystems unparalleled elsewhere in lowland Britain. Figure 5.3.2 shows the range of vegetation that exists within the **perambulation** of the Forest. Management of the New Forest is largely the responsibility of the Forestry Commission.

It has a statutory duty, under the Forestry Act of 1967, to supervise the development of afforestation and the production and supply of timber. Wide-ranging discussions took place in the late 1980s concerning the future status of the New Forest. At one stage it seemed possible that the Forest

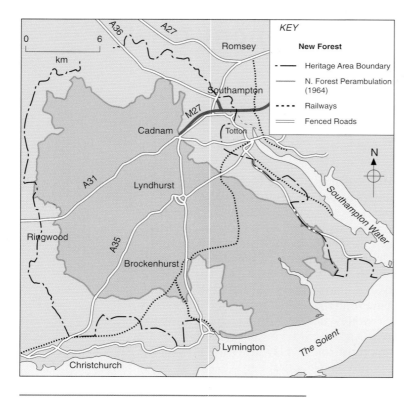

Figure 5.3.1 *New Forest Heritage Area and Perambulation*

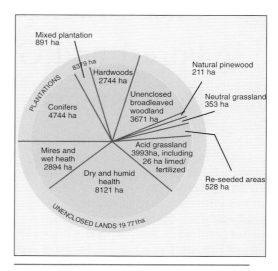

Figure 5.3.2 *Vegetation types in the New Forest*

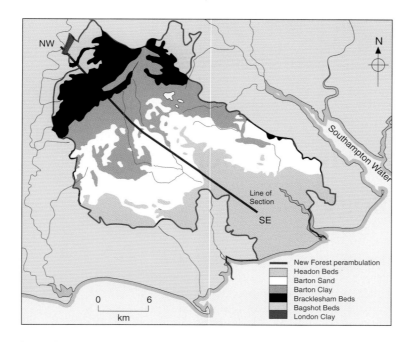

Figure 5.3.3. *New Forest: Simplified solid geology*

might become a National Park, but the final decision of the Review Group that had been examining its status was that the New Forest did not need the administrative organisation of a National Park. Instead it recommended the

establishment of a Heritage Area, somewhat larger than the existing area (see Figure 5.3.1) administered by the Forestry Commission, which should be recognised by the Government as requiring special protection because of its national and international importance. Management and planning in the Forest is today co-ordinated by a Heritage Committee under an independent chairman.

STUDENT **5.1** ACTIVITY

1 Use a good atlas to consider how the New Forest differs in location from the National Parks of England and Wales.
2 Why do you think that the Review Committee came to the conclusion that it did not need the administrative organisation of a National Park?

THE PHYSICAL BACKGROUND

Figure 5.3.3 shows a simplified geological map of the New Forest, and Figure 5.3.4 shows a cross section through the New Forest, relating the main types of vegetation to the underlying geology. The New Forest lies in the centre of the Hampshire Basin, and its geology consists simply of sands and clays overlain by gravel. The cross section shows that the gravel cappings are most extensive in the north west and south east of the Forest, and they generally coincide with rather acid **podzolic** soils, which encourage

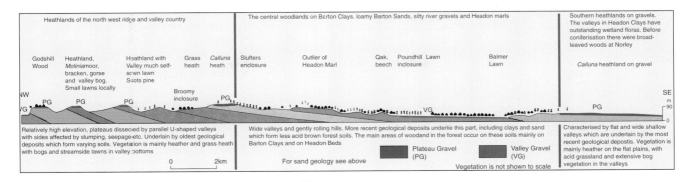

| Heathlands of the north west ridge and valley country | The central woodlands on Barton Clays, loamy Barton Sands, silty river gravels and Headon marls | Southern heathlands on gravels. The valleys in Headon Clays have outstanding wetland floras. Before coniferisation there were broad-leaved woods at Norley |

Godshill Wood — Heathland, *Molinia* moor, bracken, gorse and valley bog. Small lawns locally — Heathland with valley much self-sown lawn Scots pine — Grass heath — *Calluna* heath — Broomy inclosure — Slufters enclosure — Outlier of Headon Marl — Oak, beech — Poundhill inclosure — Lawn — Balmer Lawn — *Calluna* heathland on gravel

NW   SE

Relatively high elevation, plateaus dissected by parallel U-shaped valleys with sides affected by slumping, seepage etc. Underlain by oldest geological deposits which form varying soils. Vegetation is mainly heather and grass heath with bogs and streamside lawns in valley bottoms.

Wide valleys and gently rolling hills. More recent geological deposits underlie this part, including clays and sand which form less acid brown forest soils. The main areas of woodland in the forest occur on these soils mainly on Barton Clays and on Headon Beds
For sand geology see above

Plateau Gravel (PG)   Valley Gravel (VG)
Vegetation is not shown to scale

Characterised by flat and wide shallow valleys which are underlain by the most recent geological depostis. Vegetation is mainly heather on the flat plains, with acid grassland and extensive bog vegetation in the valleys

0   2km

Figure 5.3.4 *New Forest: Geology, landform and vegetation*

Figure 5.3.5 *Heathland in the New Forest*

heathland (see Figure 5.3.5). In the central area, the lower Valley Gravels are more silty, and, together with the sands and clays present here, give rise to much more fertile brown forest soils. It is in this central zone that the main woodlands (see Figure 5.3.6), which include both the Ancient and Ornamental Woodlands, and the Timber Inclosures, occur. In the north, deep valleys are incised into the gravel plateaux, and their vegetation of wet heath, acid grassland and woodland contrast with the broad sweep of intervening heathland on the gravel spreads. Much shallower valleys are characteristic of the southern part of the Forest, and their peat accumulations encourage the development of bog vegetation.

ECOSYSTEMS OF THE NEW FOREST

The complex pattern of vegetation in the New Forest has developed over a period of some 13 000 years since late glacial times. At that time the New Forest would have experienced a periglacial climate, and would have been covered with tundra

Figure 5.3.6a *Ancient and Ornamental woodland, Rhinefield Drive, New Forest*

Figure 5.3.6b *Timber Inclosure, Wilverley, New Forest*

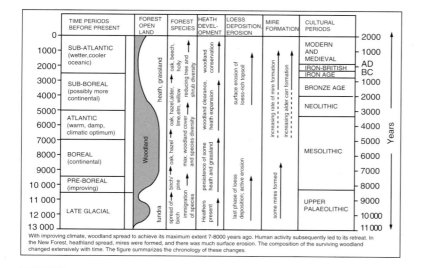

With improving climate, woodland spread to achieve its maximum extent 7-8000 years ago. Human activity subsequently led to its retreat. In the New Forest, heathland spread, mires were formed, and there was much surface erosion. The composition of the surviving woodland changed extensively with time. The figure summarizes the chronology of these changes.

Figure 5.3.7 *Evolution of ecosystems in the New Forest*

vegetation. Evolution of the present ecosystems from that period is shown in the diagram Figure 5.3.7.

STUDENT **5.2** ACTIVITY

1 Use Figure 5.3.7 in conjunction with Figure 5.3.8 which shows the present distribution of vegetation in the New Forest
   a Carefully trace the evolving relationship between forest and open land in the period between the late glacial period and the present.
   b What appear to be the main physical and human factors responsible for this changing relationship?

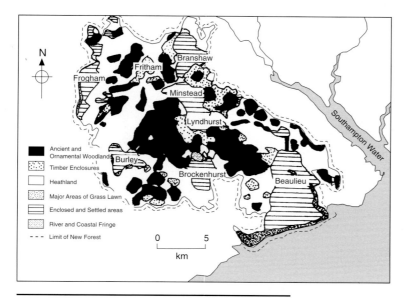

Figure 5.3.8 *New Forest: distribution of vegetation types*

c What would have been the main factors responsible for
   i the loss of diversity in trees and shrubs in the later phases
   ii the increasing degree of mire formation in the later stages?
2 Relate the relative amounts of vegetation present today (as shown in Figure 5.3.8) to the evolutionary pattern as shown in Figure 5.3.7).

THE FOREST ECOSYSTEM AND ITS MANAGEMENT

It will be noted from Figure 5.3.8 that there is a clear distinction between the unenclosed woodlands (3882 hectares) and the Forest plantations in the Inclosures (8513 hectares). The former are the Ancient and Ornamental woodland, which are the present day remnants of the ancient pasture woodland, whilst the plantations (or Inclosures) represent the areas which concentrate on present day timber production.

THE ANCIENT AND ORNAMENTAL WOODLANDS

The main features of the structure and functioning of the ecosystem of the Ancient and Ornamental woodlands are shown in Figure 5.3.9. These woodlands are the oldest part of the forest, and date back to the time when it was declared a royal forest in the eleventh century. It was initially established as a deer forest, and its role as pasture for animals, such as deer, cattle and ponies has remained important until the present. Management practices in the past have had some important effects on the present structure of the forest:
- the continual use of the forest for pasture has resulted in a poorly developed herb layer;
- **coppicing** of hazel resulted in its retention in the woodland system, but as the practice died out, hazel began to disappear from the forest, and has been largely replaced by holly as the main tree in the shrub layer;
- **pollarding** produced firewood and browse for deer, but was prohibited by the 1698 Inclosure Act, largely on the grounds that it did not encourage timber production. Pollarded trees that remain today add character to the Forest,

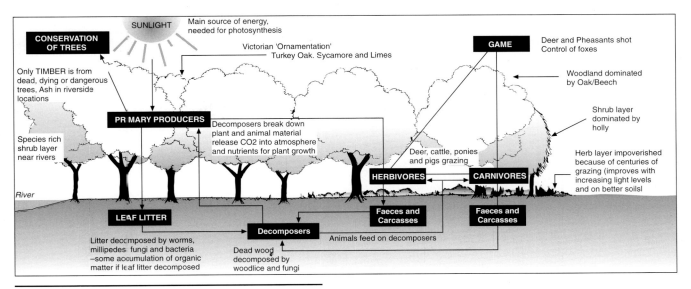

Figure 5.3.9 *Ecosystems of Ancient and Ornamental woodlands*

although, because of their great age, they are unlikely to be a permanent feature of the woodlands;

- exploitation of oak for naval timber from the early seventeenth to the nineteenth century has tended to adjust the species composition within the Ancient and Ornamental woodlands towards beech (in 1707 the ratio of oak to beech was 3.2 : 1, but a relatively recent survey showed that it is now 1 : 1);
- Victorian ornamentation of the woods led to the establishment of a range of exotic species such as maples and conifers;
- 'silvicultural management' in the period between 1950 and 1970 led to the loss of uneven aged woodland, the break up of the canopy, and damage to soil structure.

Management of the Ancient and Ornamental woodlands is today concerned with maintaining its traditional character, and there is therefore a general policy of non-intervention. Conservation is a key priority, and English Nature is fully consulted on all matters connected with conservation. Current management in the Ancient and Ornamental woodlands also includes some provision for regeneration. Although nearly three quarters of the broadleaved high forest possesses adequate successor crops, some of the remaining areas are severely impoverished due to high levels of grazing and tree deaths. Thus small inclosures may be made within these areas, together with thinning of advanced regeneration and artificial plots.

STUDENT 5.3 ACTIVITY

1 Briefly examine the results on the ecology of the Ancient and Ornamental Woodlands of:
  a an increase in the number of ponies and cattle grazing;
  b the enclosing of certain areas so that no grazing is permitted.
2 Figures for the last forty years show a considerable year-on-year fluctuation in the number of animals grazing in the Forest. Suggest why these figures need to be closely monitored.

THE TIMBER INCLOSURES

The main Timber Inclosures are shown in Figure 5.3.8. Three main types of Inclosure exist and their relative areas are shown below:

Statutory Inclosures   7107 hectares
Verderers' Inclosures   741 hectares
Crown Freehold Inclosures   470 hectares.

Statutory Inclosures were planted as a result of Acts of Parliament in the late eighteenth century and the nineteenth century. Many of these plantations have been felled in the twentieth century, and have been replanted with conifers. The Verderers' Inclosures were planted on open heathland in the late 1950s and early 1960s. The pattern of planting from the years before 1790 to the present day is shown in Figure 5.3.10 and the species composition of the Inclosures is shown in Figure 5.3.11.

Planting Year	Age Class (years)	Area of Conifer (ha)	Area of Broadleaves (ha)
1981–1990	10	301.4	29.8
1971–1980	20	333.3	8.0
1961–1970	30	1370.3	72.0
1951–1960	40	660.6	108.6
1941–1950	50	1195.9	233.3
1931–1940	60	175.1	162.3
1921–1930	70	434.7	147.8
1911–1920	80	69.2	59.6
1901–1910	90	3.9	33.6
1891–1900	100	7.8	26.7
1881–1890	110	–	5.3
1871–1880	120	0.8	23.3
1861–1870	130	22.6	246.7
1851–1860	140	22.1	540.8
1841–1850	150	0.7	84.5
1831–1840	160	–	4.2
1821–1830	170	–	157.9
1811–1820	180	–	260.8
1801–1810	190	1.9	459.9
1791–1800	200	–	9.1
Pre 1790	+200	–	36.6

Figure 5.3.10 *Pattern of planting in inclosures*

	Approximate % of plantation area
CONIFER HIGH FOREST	
Scots Pine	21
Corsican Pine	13
Douglas Fir	12
Spruce	6
Larch	2
Other	6
TOTAL	60
BROADLEAVED HIGH FOREST	
Pure Oak	12
Mixed Oak/Beech	18
Pure Beech	3
Potential BHF	6
Other	1
	40

Figure 5.3.11 *Species composition of the Inclosure Woodland*

STUDENT 5.4 ACTIVITY

1 Comment on the pattern of planting as revealed in Figure 5.3.10.
2 Discuss some of the factors that will have influenced the mix of species shown in Figure 5.3.11.

The structure and functioning of the ecosystem in the coniferous inclosures is shown in Figure 5.3.12.

STUDENT 5.5 ACTIVITY

1 Make a careful comparison of the structure and functioning of the ecosystem in the Ancient and Ornamental Woodlands with that in the Coniferous Inclosures. Summarise your conclusions in a table.

Most of the conifer plantations derive from replanting of felled conifers and the conversion of broad-leaved woods after 1920 and between 1940 and 1970. They also include all of the conifers from the relatively recent Verderers' plantations. Conversion to conifers from broad-leaved woods ceased in 1970. A rotation of 200 years is required for all broad-leaved woods. Current restocking takes place at a rate of 30 to 40 hectares a year. Current timber production is of the order of 40 000 m³ a year. The management objectives for both broad-leaves and conifers are shown in Figure 5.3.13.

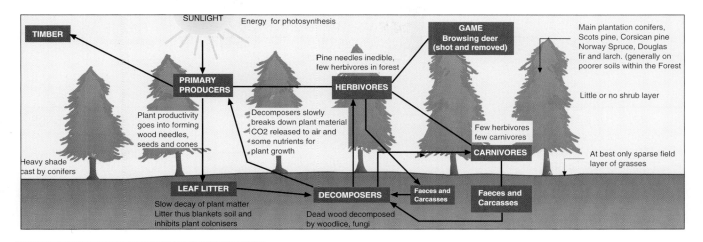

Figure 5.3.12 *Ecosystems in Coniferous Inclosures*

STUDENT   **5.6**   ACTIVITY

1 What are the essential differences
between the management objectives for
broad-leaves and conifers in the New
Forest Inclosures?
2 Explain the reasons for these differences.

BROAD MANAGEMENT ISSUES IN THE NEW FOREST

Management issues considered so far have
been largely confined to the forest areas
themselves, i.e. the Ancient and
Ornamental Woodlands and the Timber

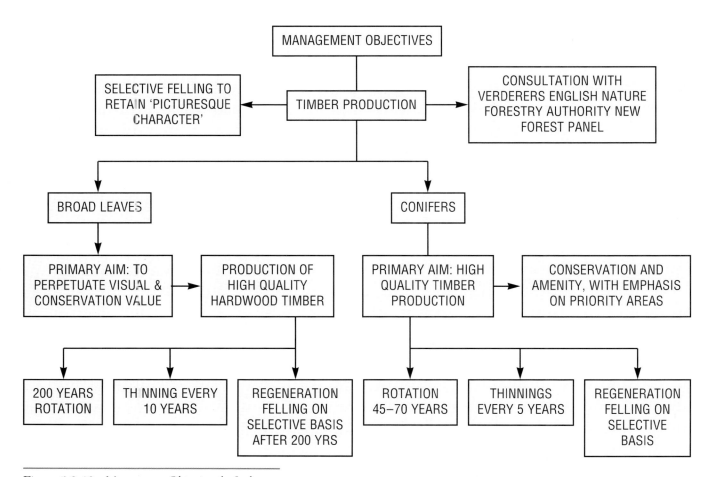

Figure 5.3.13 *Management Objectives for Inclosures*

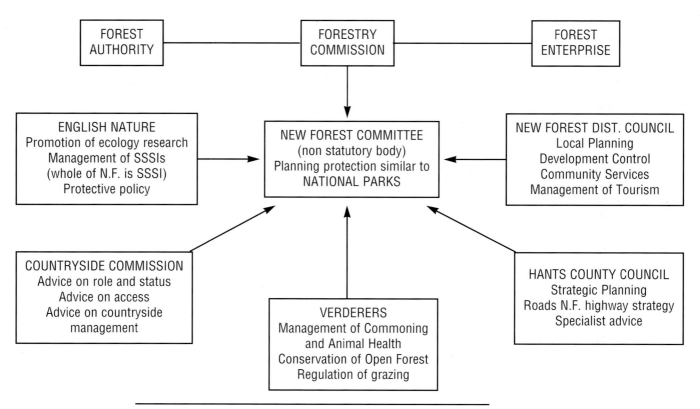

Figure 5.3.14 *Relationship between Forestry Commission and other bodies*

Inclosures. The Forestry Commission, however, has much broader responsibilities within the New Forest perambulation, and in the more recently designated Heritage Area. In discharging these duties it has to work with a number of other bodies that also have responsibilities within the Forest. Figure 5.3.14 shows the relationship between the Forestry Commission and these other bodies.

Apart from its supervision of woodland management, the Forestry Commission has important responsibilities in a number of other fields in the New Forest:

- the maintenance of open forest ecosystem which will provide adequate grazing for Commoners' animals;
- preserving the high quality landscape of the New Forest so that its traditional character is maintained;
- maintaining and protecting the ecosystems within the New Forest that have a high conservation status (the whole of the New Forest is a Site of

Special Scientific Interest and has a status equivalent to a National Nature Reserve);
- providing facilities for a wide range of leisure and recreational activities;
- monitoring of leisure and recreational activities that could have harmful effects on the general well-being of the Forest.

S T U D E N T ● **5.7** A C T I V I T Y

1 Indicate how management duties for the above have to be shared between the Forestry Commission and the other bodies. You may wish to summarise your ideas in a flow diagram.
2 Some of the objectives in the broader management of the New Forest may lead to a conflict of interests. Draw a conflict matrix to show where such conflicts may arise, and suggest how compromises might be achieved.

5.4 The Galloway Forest Park

This Forest Park lies on the west of the Southern Uplands of Scotland, and illustrates well the management issues associated with the establishment of coniferous plantations in an upland environment. Considerable variety exists within the area, with marked contrasts existing between the relatively sheltered broadleaved woodlands in Glen Trool, the uniform stands of Sitka spruce on the slopes of Kirrierioch Hill, and the wilderness areas that remain unplanted in the central moorland areas on the flanks of The Merrick. Although timber production is the prime reason for the afforestation of these uplands, management has inevitably become involved with other issues, such as conservation, landscape design and recreation.

The map (Figure 5.4.1) shows the extent of afforestation in Galloway, and the various stages in its development. Initial land was acquired for planting by the Forestry Commission in the present Kirroughtree and Bennan forests in 1921. Despite these early beginnings most of the afforestation has taken place since World War II. The choice of trees that can be planted is much restricted by the harsh climatic conditions and the poor soils that exist over much of the area. The table, (Figure 5.4.2), shows the main species that can be considered for growing in the Scottish Uplands, and the conditions under which they flourish. The composition of the plantations in the Newton Stewart District of the Galloway Forest Park in 1994, and those proposed for 2014 are shown in Figure 5.4.3. By way of comparison the equivalent figures for Britain are given.

STUDENT ● 5.8 ACTIVITY

1 Using the tables (Figures 5.4.2 and 5.4.3) examine the ways in which the composition of the Newton Stewart plantations reflects the physical conditions of the area.

2 The proposals for 2014 show some important differences. Identify these differences, and suggest some possible reasons for them.

SENSITIVITY AND THE LANDSCAPE ISSUE

In recent years there has been much discussion concerning the need for good landscape design in upland afforestation. In the past, afforestation blocks were of the order of 20–30 hectares, and were usually set out in rectilinear units that often formed an alien limit in the landscape. As a result the Forestry Commission has come in for a great deal of criticism over its planting policies, and their effect on the upland landscape.

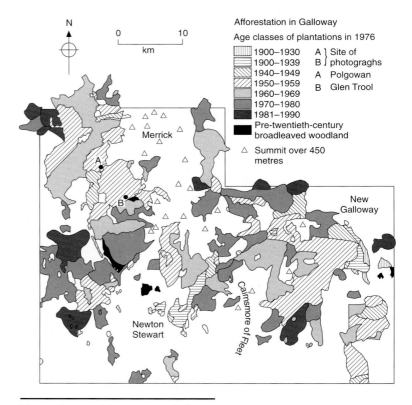

Figure 5.4.1 *Afforestation in Galloway*

Tree species	Soil	Climate	Altitude	Timber
Scots Pine	Light, sandy soil. Avoid waterlogged areas	Does well in low rainfall areas. Very frost hardy	Does not do well at high elevations	Growth slow: volume production not high
Corsican Pine	Sandy soils near sea	Low rainfall areas	Avoid high elevation. Not suitable for N & W uplands of Britain	Timber producer faster than Scots Pine
Lodgepole Pine	Poor soils, sand dunes and peat	Stands exposure well	A good pioneer species at higher altitudes	Similar to Scots Pine
European Larch	Does best on moist well drained fertile loams. Avoid peat soils	Does not do well in exposed sites	Avoid exposed sites at high altitude	Heavy, strong timber
Japanese Larch	Avoid dry sites	Does well in high rainfall areas	Avoid very exposed sites but generally suitable for upland sites	Strong timber
Douglas Fir	Well drained soil, good depth and moderate fertility	Liable to windthrow on soft, wet ground	Unsuitable for exposed positions. Best on valley slopes	Excellent constructional timber
Norway Spruce	Most soils of moderate fertility	Does poorly on dry sites, poorly on E side of Britain	Somewhat sensitive to exposure	Good, general purpose timber
Sitka Spruce	Damp sites: avoid all dry sites	Very suitable for high rainfall areas on W coast	Stands exposure better than any other conifer	First class pulpwood, suitable for chipboard

Figure 5.4.2   *Suitability of Conifer Species for Afforestation*

Tree species	Newton Stewart District		Great Britain
	1994	**2014**	
Sitka Spruce	60	58	40
Norway Spruce	14	13	9
Larches	15	17	11
Douglas Fir	1	2	4
Scots Pine	1	1	18
Other conifers	1	–	2
Broadleaves	2	8	n/a
Open space	7	18	n/a

Figure 5.4.3   *Composition of Plantations, Newton Stewart and Great Britain*

The concept of sensitivity has recently been developed to provide guidelines for future developments. Four levels of sensitivity are recognised.
Level one: forests that are not visible or forming only long distance viewing opportunities from railways, public rights of way, minor roads and minor communities. *Level two:* highly visible and dominant forest landscapes as viewed from railways, public rights of way, minor roads and minor communities.

Figure 5.4.4 *Forest plantations, near Polgowan seen from a minor road, Glen Trool Forest District, Southern Uplands, Scotland*

Level three: highly visible and dominant forest landscapes with sensitive views from major public highways, communities, tourist routes and major Forestry Commission recreational facilities.
Level four: wildlife and archaeological sites scheduled in conservation plans.

Figure 5.4.5 *Deciduous and coniferous forest, Glen Trool, major recreation site*

STUDENT ACTIVITY

1 Study the two photographs (Figure 5.4.4 and 5.4.5). Their locations are shown on the map, (Figure 5.4.1).
2 Suggest what sensitivity levels would be appropriate for the two sites, giving your reasons.
3 Environmental objectives for the two sites may be selected from the following list;
    ● Landscaping
    ● Conservation of wildlife
    ● Recreation.
   For each of the sites select the suitable environmental objective. You may consider that more than one would be appropriate: in that case, rank your selections in order of priority.

LONG TERM MANAGEMENT OBJECTIVES

In the Newton Stewart District of the Galloway Forest Park seven long term objectives have been identified:

● Commercial timber
● Landscape design
● Diversification of age structure and species composition
● Enhancement of wildlife
● Provision of high quality recreational facilities
● Reduction of acreage of evergreens above 330 m
● Improvement of forest design around farmland within the main forest areas.

STUDENT ACTIVITY

1 Rank these objectives in order of priority. You may need to develop a matrix in order to compare their relative impacts on various aspects of the forest environment.
2 For the top three rankings in your scheme suggest what important guidelines would need to be established in order to achieve a satisfactory environmental and ecological balance between the top three and those objectives with a lower priority.

ESSAY

1 To what extent have environmental considerations been important in changing forestry management practice?

FIELDWORK OPPORTUNITIES & PROJECT SUGGESTIONS

1 Select a small piece of woodland, or a manageable tract of a larger forest area. Carry out a thorough ecological survey to establish the species structure of the woodland and the nature and depth of the soil, and the ambient conditions within the woodland (light, windspeed, temperature).
   Examine the way in which the woodland is managed, as a timber resource, and as a recreational resource. You will need to contact the owner or manager of the woodland, and to consult the management plan. Then you can plot on a map the different management initiatives for both timber production (if there is any) and recreational use.
2 Investigate the way in which a local nature reserve is managed. You should map its extent, and then carry out sample ecological surveys to show the difference between the different plant associations. It will be important to interview the warden, who can tell you about the different management initiatives that are used to maintain the status of the reserve.

MANAGING HUMAN ENVIRONMENTS

6

Managing Urban Environments

KEY IDEAS

- Processes of change and development in urban areas need to be managed in order to achieve the optimum functioning of the urban system
- Quality of life varies enormously within individual cities and towns, and a prime objective of management is its improvement in underprivileged and deprived parts of urban areas
- Service provision in towns and cities often requires skilful management of limited resources to ensure fair and equitable distribution
- Management of urban growth has to balance controlled development with the maintenance of a healthy urban environment

6.1 Introduction

Britain's population is overwhelmingly urban, and lives in an environment that presents enormous management challenges. Although Northern Ireland and the Republic of Ireland show smaller concentrations of their populations in the urban environment, it is still important to remember that roughly a third of the population in Northern Ireland lives in Belfast, and a similar proportion of the Republic's population lives in Dublin. Management of the urban environment has traditionally been in the hands of the town and city councils in both Britain and Ireland. However, these councils have to operate within broad guidelines laid down by central governments in such important fields as housing, environment and transport.

In the last three decades a whole range of important initiatives have flowed from governments only too aware of the general decline in urban environments. However,

much of the focus has been on the 'inner city', since this was perceived as that part of the urban environment where the problems were most serious. In both Britain and Ireland this was not necessarily true, because, apart from the inner areas of decay, some of the worst conditions existed on the housing estates which are located on the periphery of towns and cities.

If inner city initiatives dominated much of the political agenda of the 1970s and 1980s (Mrs Thatcher's famous clarion call on Election Night, 1987, was 'On Monday we have a big job to do in some of those inner cities'), other issues have tended to surface in the current decade. One writer, David Banister has noted that the 1980s were very much the decade of the car, with a 40 per cent increase in traffic, and a 30 per cent increase in the numbers of cars and taxis. As a consequence city centres became increasingly crowded with traffic, and retailers sought relief from these conditions by switching the focus of their new operations to city fringe locations. Such was the attraction and success of these new out-of-town centres that serious concern began to be expressed in the early 1990s for the well-being of the city centre. Thus urban managers were faced not only with the on-going problems of the inner city and the outer city estates, but also with the prospect of an irreversible decline in city centre shopping, and the growth of a range of new land-use problems on the city margins.

Significantly, important new guidelines have come from the British Government on out-of-town shopping centres in the early 1990s. Faced with city centre shopping decline which might reach the scale of that in some cities in the United States, local authorities are now actively encouraged by central government to

refuse planning permission for out-of-town centres. This would apply not only to shopping centres, but also office developments and leisure centres and parks.

STUDENT **6.1** ACTIVITY

1 Study the newspaper article 'Call for tougher action to cut urban sprawl' (Figure 6.1.1).
   a Identify the main urban management issues mentioned in the article.
   b Explain why the researchers quoted in the article do not regard present policies as adequate.

Within the city, urban managers face other problem issues that stem from the growth in traffic. Faced with increasing congestion, most large cities have explored the

Call for tougher action to cut urban sprawl

BRITAIN'S cities and towns will become increasingly sprawling and suburban in character, in spite of recent government policies intended to safeguard remaining countryside from development, two researchers warned yesterday at the Royal Geographical Society's annual conference in Glasgow.

The environmentalists' and the Government's shared desire to have compact high-density cities – more European than American in character – with highly efficient, well-used public transport systems squeezing out the private car, would not be realised without far tougher policies, Michael Breheny, a town planner, and Ian Gordon, a geographer, both of Reading University, said.

The move out of town that has already taken place, growing reliance on the private car and people's demand for more personal space as they became more affluent were all spreading Britain's urban areas more thinly, they said.

John Gummer, Secretary of State for the Environment, has said he wants to halt the spread of out-of-town shopping centres. Last year the Government produced planning guidance for local councils which asks them to refuse permission for large new commercial and leisure developments which can only be reached by private car.

His overall aim is to plan towns which cut the need to travel by private car to work, shop and play, thereby controlling pollution and congestion.

At the same time, the Government is wrestling with the problem of where to house more than 3 million new households expected to form between now and 2011. In last year's housing White Paper, the Government said that by 2005 half of all new homes should be built on reused land. "All these policies are well-meaning, even radical," Professor Breheny said. "But they will make little difference without deterring private car use more directly."

The researchers say there is a strong case for building new houses on derelict sites. For these tend to be in the big conurbations which are often suffering population decline and where the social infrastructure of schools, shops and hospitals already exists.

But the demand for new homes is often low in these areas and they are the last places where commercial house-builders want to build. Instead, demand is at its highest in smaller towns where there are few derelict sites and intense local opposition to new homes in the countryside or on playing fields and allotments. Places like Reading and Milton Keynes have had very rapid household growth over the past quarter-century and the children who grew up in them will fuel more growth for the next few decades.

"For some counties like Cambridgeshire, it probably will be an unfolding nightmare," Professor Breheny said. The researchers believe that if the target for building on derelict sites is to be hit there will have to be strong incentives for house-builders. "They will resist to the last," Professor Breheny said. Having examined data on fuel prices and urban densities in 32 cities around the world, they also believe higher petrol taxes can achieve far more than town planning.

"The planning system can only deliver limited things very slowly. Doubling the price of petrol could have a much greater effect," they conclude.

The Independent 5 January 1996

Figure 6.1.1   Urban sprawl: action for the future?

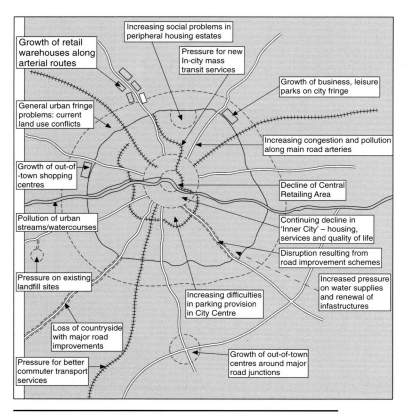

Growth of retail warehouses along arterial routes

Increasing social problems in peripheral housing estates

Pressure for new In-city mass transit services

Growth of business, leisure parks on city fringe

General urban fringe problems: current land use conflicts

Increasing congestion and pollution along main road arteries

Growth of out-of-town shopping centres

Decline of Central Retailing Area

Pollution of urban streams/watercourses

Continuing decline in 'Inner City' – housing, services and quality of life

Disruption resulting from road improvement schemes

Pressure on existing landfill sites

Increased pressure on water supplies and renewal of infastructures

Increasing difficulties in parking provision in City Centre

Loss of countryside with major road improvements

Growth of out-of-town centres around major road junctions

Pressure for better commuter transport services

Figure 6.1.2 *Major management problems and issues in urban areas*

possibility of developing some mass transit system, on the lines of those already developed in cities such as Manchester and Sheffield. Central Government now openly supports 'the implementation of measures to discourage car use' – a far cry from the 'great car economy' of the Thatcher years. Charging drivers to enter congested city centres (congestion charging) was examined by the Government, and a report published in 1995. It concluded that such a policy could not be introduced in this century, but that on-going investigations should continue. Pollution levels along arterial routes in cities have now risen to alarming levels, and the increasing occurrence of smog, particularly during periods of high atmospheric pressure, may well be related to the increasing incidence of asthmatic conditions amongst urban populations in Britain (see Chapter 9, page 168).

The diagram (Figure 6 1.2) shows the spatial distribution of the major management problems and issues in the city in Britain and Ireland in the years approaching the millennium.

6.2 Case Studies of the Management of Urban Environments

Three case studies are chosen to illustrate the management of urban environments. It was noted in the introduction to this chapter that Dublin has a very high degree of **primacy** in the Republic of Ireland with 29 per cent of the Irish population living in Dublin in 1991. Dublin well illustrates many of the management issues that face urban planners in the 1990s. In common with many other 'million cities' in Western Europe it faces a range of challenges in its inner city areas which include the renewal of areas such as Liberties, and the urgent conservation issues in Georgian Dublin both north and south of the Liffey. Although city centre retailing has remained strong in the last 20 years it has had to face increasing competition from suburban shopping centres. It has to provide housing for a rapidly growing population, which has more than doubled in the period since the end of World War II. This has been achieved in part by the development of the new towns on the western margins of the city. With such a concentration of Irish population in the city, traffic problems are serious, particularly in those streets that focus on the Liffey crossings. From a wide hinterland, the Dublin Area Rapid Transit (DART) carries huge numbers of commuters into the city each day.

One of the main management issues in any large city is the provision of housing. Although Edinburgh is not typical of Scotland as a whole, since it has a much higher proportion of owner occupiers, the City authorities still have an important role to play as a provider of housing for nearly a fifth of households. In particular

Edinburgh has to face severe social problems in housing areas such as Wester Hailes, Craigmillar and Muirhouse, all, significantly, on the periphery of the city. New towns, as a management response to overcrowding in inner city areas, and as an alternative to continued urban growth at the city margins, were characteristic of policy in Britain in the 1950s and 1960s. The example of Craigavon (designated in 1963) in Northern Ireland embodied the new principle of a corridor of development between Portadown and Lurgan, with Brownlow, the central component in the corridor being developed on a greenfield site.

6.3 Dublin: Primate City of Ireland

Joyce's image of early twentieth century Dublin is still appropriate today. Central Dublin still gives the impression of being crowded (Figure 6.3.1) and this no doubt reflects the central role that it plays in Irish life, and its primate function. As the Republic's main urban centre, with a population of over a million it is nearly ten times as large as Cork (127 000) and nearly twenty times larger than the third and fourth cities by population, Limerick (52 000) and Galway (50 000). Its primacy is reflected in the way its functions dominate Ireland as a whole:

- as the state capital, it concentrates most of the Government departments and state agencies. There has been relatively little decentralisation of the decision making function, although a few state organisations have moved out. Bord na Mona, the state-run Peat Board, has recently moved from crowded premises in southern Dublin to Newbridge, some 40 km to the south west;
- with an expected 75 per cent of its workforce engaged in **tertiary activities**, it dominates Ireland to the extent that 50 per cent of the service employment of the country is concentrated in Dublin;
- although Dublin's share of manufacturing employment has fallen quite sharply in the second half of the twentieth century (in the early 1990s it possessed barely a quarter of the country's industrial workers), it still remains the single most important industrial centre;
- it remains the unquestioned cultural centre of the Irish Republic. It has sustained its envied concentration of scholars, writers and philosophers, and can claim pre-eminence in the fields of music and the visual arts;
- it is the concentrated focus of the national and international transport systems. Nearly three quarters of the Republic's air passengers pass through Dublin Airport, and it handles two-thirds of the air freight.

STUDENT ACTIVITY

1 Dublin only possesses one EU institution (the European Foundation for the Improvement of Living and Working Conditions). Suggest reasons for this relatively poor representation.
2 Explain the apparent anomaly of Dublin's increasing dominance of the service industries of the Irish Republic, and its decreasing share of the country's manufacturing employment.

Figure 6.3.1 *O'Connell Bridge and Street, Dublin*

The grey warm evening of August had descended upon the city and a mild warm air, a memory of summer, circulated in the streets. The streets, shuttered for the repose of Sunday, swarmed with a gaily coloured crowd. Like illuminated pearls the lamps shone from the summits of their tall poles upon the living texture below, which, changing shape and hue increasingly, sent up into the warm grey evening air an unchanging, unceasing murmur.

James Joyce: The Dubliners

	City	Suburbs	Total	Sub-Region	Change (%)
1821	244 000				
1831	232 362	32 954	265 316		
1841	232 726	48 480	281 206		
1851	285 369	59 468	317 837	405 147	
1861	254 808	70 323	325 131	410 252	+1.3
1871	246 326	83 410	329 736	405 262	−1.2
1881	249 602	95 450	345 052	418 910	+3.4
1981	245 001	102 911	347 216	419 216	+0.7
1901 (old)	260 035				
1901 (new)	290 638	90 854	381 492	448 206	+6.9
1911	304 802	99 690	404 492	477 196	+6.5
1926				505 654	+6.0*
1936				586 925	+16.1
1946				636 193	+8.4
1951				693 022	+8.9*
1961				718 332	+3.7
1971				852 219	+18.6
1981				1 003 164	+17.7
1986				1 021 449	+1.8*
1991				1 024 429	+0.3*

Notes: the boundaries of the City were changed in 1990; * figures do not relate to a ten year period; Sub-region includes Dublin Co. Borough, County Dublin and Dun Laoghaire Borough (a suburb to the south-east of Dublin).

Figure 6.3.2   *The growth of Dublin's population*

GROWTH AND SPATIAL PATTERN OF DUBLIN'S POPULATION

Figure 6.3.2 shows the growth of Dublin's population between 1821 and 1991. It will be seen that the nineteenth century was a period of relatively modest growth, even of decline in some decades. In the twentieth century there has been a pattern of continuous growth, with the population doubling in the 50 years between 1936 and 1986, and passing the million mark for the first time in the 1980s. Nearly half of this growth occurred in the years between 1961 and 1981, with most of this growth being accounted for by natural increase. If, however, the pattern of population in Dublin City itself is compared with that of County Dublin, then some significant differences are noted.

	1961	1966	1971	1979	1981	1981[a]	1986[a]	1991[a]	
Co. Borough	537 448	567 802	567 866	544 568	525 882	544 833	502 749	477 675	
Dun Laoghaire	47 792	51 811	53 171	54 244	54 496				
						178 116	180 675	185 362	New admin
Co. Dublin	133 092	175 434	231 182	384 853	422 786	165 264	199 564	208 666	districts
						114 952	138 479	152 725	
Dublin	**718 332**	**795 047**	**852 219**	**983 665**	**1 003 164**		**1 021 467**	**1 024 429**	
Ireland	**2 818 300**		**2 978 200**		**3 443 400**		**3 540 600**	**3 523 400**	
Dublin as %	25.5%		28.6%		29.1%		28.9%	29.1%	

(a) Figures relate to the newly defined areas following the Local Government (Reorganisation) Act, 1985. Direct comparison of Census data is rendered impossible over the longer run as the population figures have not been adjusted by the Central Statistics Office for census years prior to 1981 to take account of the revised boundaries.

Figure 6.3.3   *Recent population growth in the Dublin area*

STUDENT 6.3 ACTIVITY

1 Study Figure 6.3.3 which shows population change in Dublin City, Dun Laoghaire (Dublin's port) and County Dublin.
  a Compare population change in Dublin City with that in County Dublin in the years between 1971 and 1981. What are the important differences in population change in the two areas?
  b To what extent is the same trend apparent in the period between 1981 and 1991?
  c What important processes in the evolution of the population of Greater Dublin do these figures illustrate?
  d What geographical factors would influence the operation of these processes?

The population loss from Dublin City was particularly noticeable in the inner city areas. In 1926, the inner city with a population of 269 000, accounted for nearly two-thirds of the population of Dublin City but by 1986 its population had fallen to a mere 83 200. This trend has continued into the 1990s, with a loss of 5000 inhabitants on average each year.

STUDENT 6.4 ACTIVITY

Figure 6.3.4 shows the age structure of the population of Dublin by district.
1 Comment on the demographic differences between the inner city areas, and the new towns of Tallaght, Clondalkin and Blanchardstown.
2 What demographic differences exist between the owner-occupied suburbs of Clonskeagh and Blackrock, and the areas of public housing at Crumlin (1930s) and Finglas (1960s)?
3 Attempt to establish a demographic gradient between the inner city and the outer public housing estates and new towns.

Figure 6.3.5 shows social class structure for the same sample areas within Dublin.
4 What features of social class structure distinguish the inner city areas from the new towns on the western fringe of Dublin?

5 What is the distinctive profile of occupational status in
  a the owner-occupied suburbs
  b the areas of public housing?
6 Review the overall relationships between demographic structure and occupational status as revealed in the profiles.

The overall pattern of social areas in Dublin is shown on the map (Figure 6.3.6) and an explanation of each type area, together with its percentage of total population is shown in the accompanying table (Figure 6.3.7).
7 Bearing in mind contemporary patterns of population change, suggest how the pattern of social areas in Dublin might change over the next twenty years.

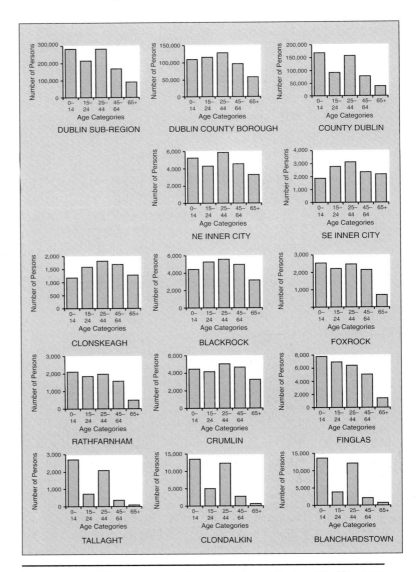

Figure 6.3.4 *The age structure of the population of districts of Dublin, 1986*

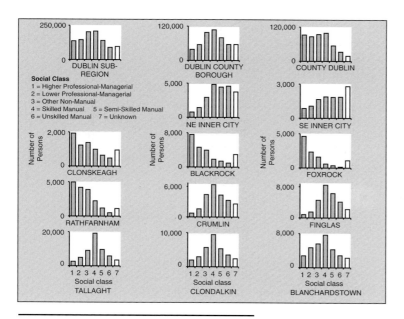

Figure 6.3.5   *Social class structure within Dublin*

HOUSING DUBLIN'S POPULATION

With rapid population growth in the suburban areas of Dublin, there is inevitably a marked contrast between the age and quality of the housing stock in the commuter belt, and the older housing in inner city areas. In suburban County Dublin, 75 per cent of the housing stock has been completed since 1961, whilst in the Inner City areas, 50 per cent of the dwellings are pre-1919. Severe overcrowding was a grim characteristic of much Inner City housing in the nineteenth century and early twentieth century, with a Parliamentary report in 1914 finding that 12 000 families were living in single room units. By 1981 this figure had been reduced to a mere 300. However, at that time 43 000 people were living at a density of two or more persons per room.

The overcrowded tenements of the nineteenth century influenced future thinking on housing provision to the extent that most suburban development has been at relatively low densities of 20–25 dwellings per hectare. However, cost considerations and developers' interpretations of planning regulations have led to a dreadful uniformity of housing in much of suburban Dublin as shown in Figure 6.3.9.

It is the expressed policy of Dublin Corporation that 'every family unit should have a decent standard of housing', and to 'seek ways and means to stimulate private housing, particularly in the Inner City'. In what is a relatively poor country by Western European standards it is perhaps surprising that, in Ireland as a whole, the level of owner occupation is as high as 80 per cent, with Dublin lagging behind with a figure of just under 70 per cent. The changing pattern of public and private sector housing development is shown in Figure 6.3.8. Sadly, the provision of public housing now seems to be confined to the needs of an underprivileged minority. Overall patterns of home ownership in Ireland and the Dublin region are shown in Figure 6.3.10.

A number of factors have combined to lead to the high levels of owner-occupation in Dublin:

- encouragement of the private housing sector by both national and local government policies;

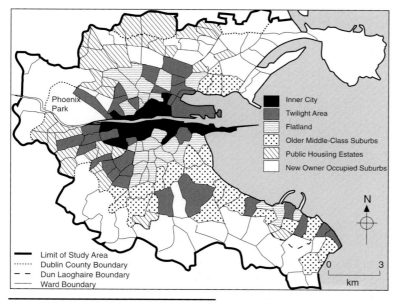

Figure 6.3.6   *Social areas in Dublin, 1971*

Social area	Characteristics	% Total population
Inner city	Low income, multiple dwellings, acute social disadvantage	11.0
Twilight areas	Areas of older housing and population, few children or professionals	15.4
Flatland	Multi-occupancy, late nineteenth century, small households and small dwellings	12.6
Old middle-class suburbs	Older middle-class population, substantial dwellings	9.7
Local authority suburbs	Local authority housing, large families, manual employees, unemployment	22.3
New owner-occupied suburbs	Young and growing populations	29.0

Figure 6.3.7 *Dublin's social areas and sub-areas*

- the expansion of the Irish building society movement from the 1960s;
- the increasing participation of Irish banks in the mortgage business (34 per cent of all new loans advanced in 1990);
- pricing policies amongst developers, particularly designed to attract first time buyers;
- the availability of public sector housing for purchase by sitting tenants (the equivalent of the '**right-to-buy**' policy in Britain);
- tax relief on mortgage interest payments;
- cash grants to first-time buyers of new dwellings, and exemption from stamp duty on the purchase of new dwellings below a certain price threshold.

Both in Ireland as a whole and in Dublin in particular the public housing sector has been recognised as something of a safety net for those who cannot aspire to owner occupation. It therefore differs markedly

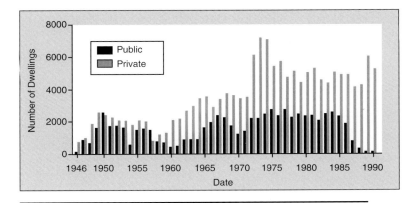

Figure 6.3.8 *Pattern of public and private housing development in Dublin 1945–90*

from Britain, where, until the Thatcher years at any rate, the public sector catered for a wide social mix of tenants.

Public sector housing in Dublin had its beginnings in the late nineteenth century and the early twentieth century, when infill and slum clearance sites in the inner city were used for the building of apartment blocks or two-storey cottages. In the inter war period flats continued to be built in the inner city area, but the public sector expanded into the development of suburban housing. The construction of flats of greater height than three storeys was more or less banned in the early 1970s and this led to a concentration of high density low rise buildings in the central areas (Figure 6.3.11). However the increasing shortage of sites in the inner city areas led to more and more public sector housing being built on greenfield sites in the outer areas, leading, inevitably to urban sprawl.

Figure 6.3.9 *Housing in Tallaght New Town, Dublin*

	1961		1981		1981		
	Dublinª	Ireland	Dublinª	Ireland	Co. Borough	Dun Laoghaire	Co. Dublinᵇ
Owner occupiedᶜ	41.6	59.8	65.6	74.4	57.5	62.3	78.1
Privately rented (unfurnished)	24.1	14.9	4.4	3.8	6.2	6.6	1.4
Privately rented (furnished)	5.9	2.3	6.3	6.3	15.5	12.4	3.7
Local Authority	25.6	18.4	17.2	12.5	18.8	16.0	15.5
Other	2.8	4.6	1.9	3.0	2.0	2.7	1.3

ª For the purposes of comparability between the Census of 1961 and 1981, the term Dublin refers to Dublin County Borough, Dun Laoghaire Borough and all of County Dublin.
ᵇ The 1981 figures for County Dublin refer to the county's aggregate town areas.
ᶜ Including tenant purchase and vested cottages schemes.

Figure 6.3.10   *Housing tenure*

It is the policy of Dublin Corporation to seek the redevelopment and rehabilitation of the Inner City.

Dublin City Plan 1991.

New starts in public sector housing were badly hit by financial restraints, and by the early 1990s they had virtually stopped. The privately rented sector has shown a steady decline in the late twentieth century. Landlords in this sector often fail to invest enough money in maintenance of their properties, which have tended to decline in number after slum clearance and redevelopment programmes in inner city locations. Tenants are usually in the lower income groups, and at a particular stage in their life cycle, where mobility, and lack of finance for long term investment in owner occupation are important motives for seeking privately rented property. Higher income groups involved in this sector clearly have mobility as a major incentive, and generally seek properties that carry a high rent. Stock in this sector varies enormously from luxury flats at the top end of the market, to small and often inadequate bedsits at the lower end.

Figure 6.3.11   *High density low-rise buildings, Inner city Dublin*

STUDENT   **6.5**   ACTIVITY

1 Discuss the main factors responsible for present patterns of tenure of housing in Dublin.
2 Bearing in mind current trends, suggest what changes are likely to take place in the next twenty years.

MANAGING THE INNER CITY

As defined in the City Development Plan, the Inner City includes all of the area between the two Canals (see Figure 6.3.12). Within this area lie the CBD, the former Mediaeval City, the heritage areas of Georgian architecture, and some residential areas. In this section it is proposed to examine some of the Government initiatives for inner city renewal, and also to review the conservation issue in Georgian Dublin. It is first necessary to consider some of the typical 'pitiless indicators' of inner city decline in Dublin:

- loss of population (see page 103), where current estimates were given as 5000 per annum);
- an ageing population (see profile page 103) – largely a residual one after younger age groups have moved out;
- loss of manufacturing jobs – it was estimated that up to 2000 jobs were being lost annually in the late 1970s as a result of recession;

- high levels of unemployment, reaching over 80 per cent in some areas;
- a high level of dependency on the social welfare system for income support;
- low levels of participation in further and higher education;
- higher than average rates of crime, although much lower than in other European capitals;
- high levels of atmospheric pollution, particularly that generated by traffic on arterial roads leading into the city centre;
- a higher than average concentration of public sector housing;
- changing land use – with the closure of many industrial units, office developments replaced industry as the most important land-use type. In one area industrial land use declined from 45 per cent of the total area in 1966 to a mere 5 per cent in 1991;
- large areas of run-down or derelict buildings. This has resulted in considerable redevelopment in certain areas, usually in the form of office blocks and car parks. Road-widening schemes have led to **planning blight** in some areas, with abandoned buildings awaiting demolition.

Inner City Renewal

The most important initiative to stimulate the involvement of private sector capital in the rehabilitation of the Inner City was the Urban Renewal Act and Finance Act of 1986. It established a package of incentives to encourage development, not only in Dublin, but also other Irish towns.

The main measures introduced were:

- full remission of rates for ten years on new buildings and on the increased value of reconstructed properties;
- capital allowances for commercial development;
- rent allowances against tax;
- construction allowances for owner occupiers, amounting to a total of 50 per cent of the building costs;
- additional allowances for the Custom House Docks development.

There were five main areas designated in Dublin, with four being located in the inner city area (see Figure 6.3.12). One of the most prestigious developments has been on the eastern side of the Inner City, where the Customs House Docks Development Authority has been

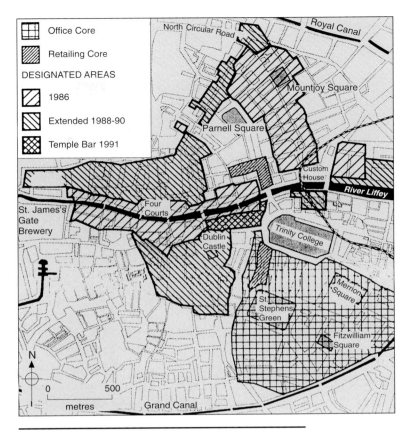

Figure 6.3.12   *Inner city designated areas for urban renewal*

responsible for the development of a 20 hectare site. Central to the development has been the creation of a new International Financial Services Centre (see Figure 6.3.13). Also included is a new national sports stadium.

Considerable progress has been made in the other designated areas, particularly on the south bank of the Liffey (see photograph, Figure 6.3.14). Much of the development has been non-residential, but some of this office space has been found to be surplus to demand. It is likely that

Figure 6.3.13   *The Custom House and the International Financial Services Centre, Dublin*

Figure 6.3.14   *Redevelopment in the Inner City, Liberties, Dublin*

residential developments will figure more prominently in future programmes.

The more recent proposals for the Temple Bar area represent a change from the previous programmes. Although the centre of this area had for some time been earmarked for the development of a central bus station by CIE (the State Transport Company), it has become an area of pubs, cafés, cheap restaurants, small theatres and galleries, much frequented by students and tourists.

In 1987 the proposal for a central bus station was rejected, and with the passage of the Temple Bar Renewal and Development Act in 1991 the area's future as a cultural and artistic centre seems assured. Proposals for the area will include a sensible mix of refurbishment with new building.

Although much of this redevelopment has had an important cosmetic effect on the inner city street façades, some question just how much it has done for inner city deprived communities. It has been noted that much of the development has been commercial rather than residential, and that there has been little in the way of permanent job creation for inner city inhabitants. New rejuvenation areas identified in the City's Development Plan of 1991 at Hanover Quay, the Mediaeval City, The Liberties and Smithfield put a much higher emphasis on residential development and refurbishment.

STUDENT 6.6   ACTIVITY

1 What are the main social problems likely to arise from the introduction of private sector capital into inner city renewal projects?
2 Rehearse the arguments for and against the Temple Bar scheme being more worthwhile than the Custom House scheme.

CONSERVATION IN GEORGIAN DUBLIN

STUDENT 6.7   ACTIVITY

Study the article from the *Guardian*, 'Fight to save Georgian Dublin' (Figure 6.3.15).
1 Identify the values positions represented by the different groups and organisations mentioned in the article.
2 Should conservation of historic buildings be valued more highly than redevelopment of derelict properties for the local community in inner city areas?

Georgian Dublin possesses some of the finest architectural heritage (see Figure 6.3.16) of any European capital city. Most of Georgian Dublin was built between 1714 and 1830, and extended over considerable areas both north and south of the Liffey. Most of the Georgian houses were built in long terraces along wide streets or around squares, such as Merrion Square in the south and Parnell Square in the north, to give an elegant and spacious urban environment. The current Development Plan of Dublin Corporation (1991) devotes much space to a number of buildings that are to be preserved. Alteration or demolition cannot occur without planning permission. Some would say that the first list of buildings worthy of preservation (in the Development Plan of 1976) appeared too late, and that irretrievable damage to the Georgian fabric of the city had already been done. One estimate suggests that 40 per cent of the Georgian buildings were lost between 1960 and 1980. Some Georgian buildings have been converted into office accommodation, and elsewhere **gentrification** has seen conversion to high class residential use. Demolition was more common in some

Fight to save Georgian Dublin

David Sharrock on a strange coalition which includes Ian Paisley and James Joyce

THE battle for Georgian Dublin has opened on another front. First it was the house of Ulster's greatest Unionist, Sir Edward Carson; now James Joyce has entered the fray.

An Taisce, the Irish Republic's equivalent of National Heritage, is locked in conflict with Dublin Corporation, accusing it of "inert, time-serving myopia" when it comes to preserving the capital's best architectural features.

The organisation is horrified at the condition of a Georgian house on Liffey quays, the setting for James Joyce's most memorable short story, The Dead. The fame of the house at 15 Ushers Island grew after John Huston set his award-winning film of the story there.

"What has happened has been horrific," said Michael Smith, chairman of An Taisce's Dublin planning group. "The front windows have been smashed, the elaborate fanlight and most of the interior woodwork stolen and the building is now just a wreck. "As well as being one of the finest surviving Georgian houses on the Liffey quays, this represents the devastating loss of a great literary landmark."

The house's owners, Heritage Properties, deny the damage is as bad as An Taisce claims, and say they will unveil restoration plans this month. "It's been a difficult building to deal with. It has taken us two years to sort out and get the plans right, but Heritage Property will be restoring it faithfully," said developer Terry Devey.

The company is in disagreement with Dublin Corporation over the original terms of the sale. "Heritage Properties were given permission to develop an adjoining site on condition they refurbish the Joyce house," Mr Smith said.

"They have blatantly failed to do so and are disputing that condition on legal grounds. We are urging Dublin Corporation to enforce its original requirement to put back the house's top storey and refurbish the interior."

A skirmish between the city fathers and An Taisce is continuing over the state of the home of Lord Carson, whose anti-home rule campaign led to the foundation of Northern Ireland. The corporation had granted three demolition applications for the house, but An Taisce enlisted the support of the Democratic Unionist leader, Ian Paisley, and the Irish government made the Carson building a national monument. Unfortunately not everybody shared the mood of peace and reconciliation. "After we highlighted the house's significance as Carson's former house, somebody broke in and smashed it up," Mr Smith said.

Another developer wants to build a hotel, demolishing the Carson monument in the process. The development has been allowed by Dublin Corporation, and after further pressure from An Taisce the Office of Public Works has stepped in.

"Our concerns are neither political nor literary, but we will use any method to bring pressure to bear on the immovable object that is Dublin Corporation," Mr Smith said.

The battle continues.

Guardian, 4 May, 1996

Figure 6.3.15   *Fight to save Georgian Dublin*

areas adjacent to the CBD, and although the new buildings were often sympathetically designed to blend in with the remaining Georgian structures, this was not always the case. Preservation of Georgian Dublin remains a long term planning problem to which there is no easy answer.

Figure 6.3.16   *Georgian Buildings, Merrion Square, Dublin*

CHANGING RETAILING PATTERNS IN DUBLIN

Retailing has been traditionally concentrated in the CBD of Dublin (see Figure 6.3.17). It extends on either side of the Liffey, from Parnell Street in the north to St Stephen's Green in the south, and from Capel Street in the west to Marlborough and Talbot Street in the east. The two principal retailing arteries are Henry Street to the north of the Liffey and the more prestigious Grafton Street to the south (see photograph, Figure 6.3.19).

In recent years, important changes have occurred in the City Centre retailing area:

- the development of in-town covered shopping centres (see Figure 6.3.18), which house a range of shops, cafés and fast-food outlets, usually with a multiple chain store or department store operating as a magnet. Five of these shopping centres now operate within the

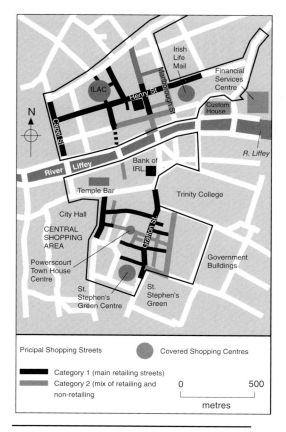

Figure 6.3.17 *Shopping streets in Central Dublin*

Figure 6.3.18 *St Stephen's Green Shopping Centre, Dublin*

CBD:
1979	Irish Life Mall	2400 m²
1981	ILAC Centre	20 400 m²
1981	Powerscourt Town House Centre	n/a
1988	Royal Hibernian Way	1580 m²
1988	St Stephen's Green Centre	21 180 m²

Figure 6.3.19 *Pedestrianised shopping, Grafton Street, Dublin*

- pedestrianisation of the main shopping streets, such as Henry Street and Grafton Street.

The increasing suburbanisation of Dublin's population has been reflected in the significant development of purpose built shopping centres in the other parts of the conurbation. In these locations these centres provide:

- a range of under-cover shopping opportunities
- a full range of shopping related services and facilities
- abundant and free car parking
- accessibility to public transport.

Figure 6.3.21 shows the proliferation of these shopping centres in the Dublin area. Although the first centres were in the 'middle' suburbs, such as Stillorgan and Phibsborough (1966), the opening of The Square, in Tallaght New Town in the 1990s (see Figure 6.3.20) initiated a new phase of development in the western suburbs.

These suburban centres clearly pose a threat to city centre locations. Suburban locations now are responsible for nearly 90 per cent of **convenience goods** sales, whilst

Figure 6.3.20   *The Square Shopping Centre, Tallaght, Dublin*

the city centre still retains a dominance in **durable goods** sales (70 per cent). Further improvements in city centre retailing locations will be needed to prevent increasing erosion of their status by the suburban centres.

Out-of-town centres have yet to make a serious impact on retailing patterns in the Dublin area. Retail warehousing is limited to furniture, electrical 'white goods', and DIY stores. The Dublin City Development Plan of 1991 will only permit a limited range of retailing opportunities in out-of-town locations. Dublin has yet to see the arrival of such developments as the Merry Hill Centre at Dudley in the West Midlands, or Lakeside at Thurrock in Essex.

STUDENT   **6.8**   ACTIVITY

1 What additional measures could be taken to make the CBD more attractive to shoppers?
2 What major planning considerations need to be taken into account before approval is given for a suburban shopping centre?

DUBLIN'S TRANSPORT SYSTEM

In common with every other major city, Dublin suffers from a number of traffic problems:

- an increasing dominance of the car in the mode of travel to work;
- traffic congestion in the central areas and on the major arterial routes leading out of the city;
- inadequate parking facilities in the city centre;

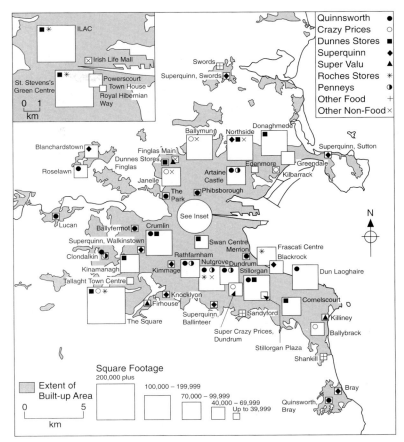

Figure 6.3.21   *Purpose built shopping centres in Dublin*

- an inadequate and expensive public transport system;
- the pursuit of policies that have encouraged the use of cars;
- the lack of a co-ordinated transport planning policy for the future.

In 1967, only 30 per cent of journeys into the city centre were by car, but by the early 1990s this had increased to well over 50 per cent. The only alternatives available were the use of the bus services, and the remaining commuter rail links (the Harcourt Street to Bray line was closed in 1958, in a suburban zone that was about to see a sizeable increase in its commuting population).

Three major reports on transportation planning were commissioned in the 1970s and early 1980s:

- the Dublin Transportation Study (DTS)
- the Dublin Rail Rapid Transit Study (DRRTS)
- the Transport Consultative Commission Report.

The main proposals of the first two reports are shown in the map (Figure 6.3.22). The

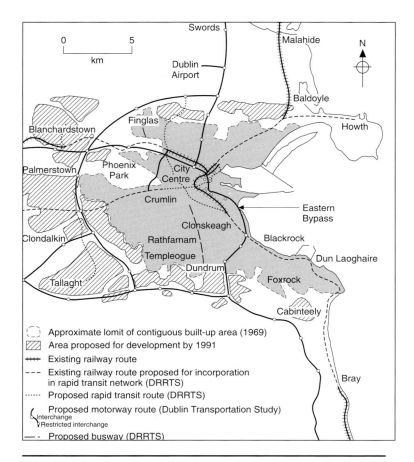

Figure 6.3.23 *DART Suburban Train, Connolly Station, Dublin*

Figure 6.3.22 *DTS and DRRTS transport planning proposals for the Dublin sub-region*

principal proposal of the DTS was for a motorway box system for the city, comprising an outer ring motorway running at about 10 km from the city centre on the west and the south, and an eastern bypass. Although it did make some recommendations for improvements in public transport, it focussed mainly on means of improving the flow of car traffic through and around the city. The DRRTS proposed over 100 km of electrified railway, linking the New Towns to a central underground station, with a similar link running north to Ballymun. It also recommended the electrification of the railway running around Dublin Bay from Howth to Bray. The final study recommended a range of low cost options to alleviate congestion in the central areas. Various recommendations in these reports have been put into operation. The western part of the motorway box was opened as the Western Parkway in 1990, and the northern and southern sections were finished in 1995. It is perhaps less likely that the eastern bypass will be completed

because of conservationist objections, although this is by no means certain. Road widening schemes were carried out in the inner areas of the city, although planning blight along their routes resulted in much abandoned and derelict property. Probably the most useful development in terms of reducing the number of cars on Dublin's roads has been DART (Dublin Area Rapid Transit) and the electrification of the coastal line from Howth to Bray (see Figure 6.3.23). Commuter trains serve the New Towns of Blanchardstown and Clondalkin by utilising existing trunk routes to Mullingar and Cork. Although the original proposals for linking Tallaght by rail to the centre were rejected on grounds of cost, they have now been revived in the 1991 Development Plan, together with proposals for a Light Rail Transit link from Tallaght to the city centre.

STRATEGIC PLANNING IN THE DUBLIN REGION

To conclude the case study of Dublin it is proposed to review briefly the main initiatives in strategic planning in the Dublin Region:

1 The Abercrombie Proposals (1941). These ideas suggested greater residential land use in the city centre, together with the creation of a Green Belt roughly 8 km in width, within which residential expansion would be restricted to growth in existing villages, which would function as satellite townships.

2 The Wright Report (1967). This report concerned itself with a much larger region, including all of County Dublin and parts of four surrounding counties. It

recommended growth being directed to several small towns outside of Dublin, including Naas, Drogheda, Navan and Arklow. Residential population in the central city was expected to fall throughout the late twentieth century. Its most far-reaching proposal was the designation of four new towns to the west of Dublin, with a projected population of 340 000 inhabitants, separated from one another and the city by green belts. In the event the four new towns were eventually reduced to three, Blanchardstown, Ronanstown (Lucan-Clondalkin) and Tallaght.

3 The Buchanan Report (1968). This was a report aimed at a national strategy for development. It affected Dublin in as far as Buchanan proposed that its growth

was to be restricted to its natural rate of increase. Eight other centres were to be actively promoted as growth poles. In the event Dublin reached a population two-thirds of the original target, whilst the other centres, with the exception of Galway, lagged some way behind.

4 The Urban Renewal Act (1986) see page 107. This was, as noted, specifically aimed at encouraging development in central Dublin and other Irish towns.

STUDENT ACTIVITY

1 Discuss some of the shortcomings of urban planning in the Dublin Region.
2 Suggest some reasons why the new towns may only be regarded as a limited success.

6.4 Managing Housing in Edinburgh

Most visitors to Edinburgh would remember it for the historic Old Town, with its Royal Mile, the spacious shopping area of Prince's Street, and the elegant town houses of the Georgian terraces of the New Town to the north. Beyond this familiar central zone lies a city that includes a wide variety of residential areas, from the inner city tenements to the terraces of Leith, and from the leafy middle class suburbs on the southern fringe to the deprived council estates on the city margins. Great differences exist in the quality of life experienced by the people living in these different urban environments. Variations in housing contribute much to these contrasts in urban living.

It has been seen that owner occupation tends to dominate housing tenure in Dublin: a similar pattern appears in Edinburgh. In this respect Edinburgh shows a striking difference from the rest of Scotland, where levels of owner occupation have traditionally been much lower. The pattern of housing tenure is shown in Figure 6.4.1. It can be noted from the diagram that there is a significant increase in the percentage of owner

occupation in the period from 1981 to 1991 from approximately 53 per cent to 67 per cent, with a corresponding decrease from 33 per cent to 19 per cent in the local authority sector. In part this reflects the continuing significance of the 'right-to-buy' policy, with over 17 000 local authority houses being sold since 1980 although there is evidence of a decline in sales in the last five years.

Although housing managers have a responsibility for housing throughout the city, available resources demand that certain areas are targeted as priority areas.

. . . there can be few cities towards which the inhabitants display a fiercer loyalty or deeper affection.

Abercrombie and Plumstead: Civic Survey and Plan for Edinburgh.

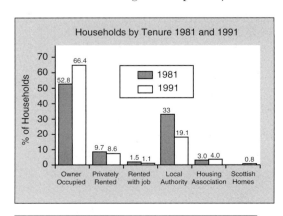

Figure 6.4.1 *Edinburgh city, households by tenure*

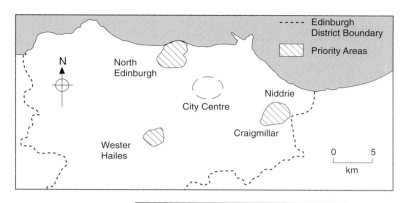

Figure 6.4.2 *Priority areas, Edinburgh City*

Figure 6.4.3a *Niddrie, Edinburgh*

Figure 6.4.3b *Wester Hailes, Edinburgh*

Figure 6.4.3c *Muirhouse, North Edinburgh*

Three such areas have been identified (see Figures 6.4.2 and 6.4.3). It will be noted that all three of these areas are located on the periphery of Edinburgh. The following demographic statistics provide a profile of the population of the three priority areas.

Area	1981	1991	% Difference
Wester Hailes	16 028	11 112	−30.6
Craigmillar/Niddrie	14 287	10 185	−28.7
North Edinburgh	30 515	24 602	−19.4
Edinburgh City	425 256	418 914	−1.5

Figure 6.4.4 *Population change*

Age	Wester Hailes %	Craigmillar/Niddrie %	N. Edinburgh %	City %
0–15	29.3	26.0	22.3	17.0
16–24	16.9	13.2	13.5	13.3
25–64	45.7	51.1	50.2	52.8
65+	8.1	9.9	14.0	16.9

Figure 6.4.5 *Age of population in priority areas*

Household type	Wester Hailes %	Craigmillar/Niddrie %	North Edinburgh %	City %
Single people	34.6	33.6	30.9	34.8
Single parents	16.9	12.9	9.6	4.1
2 Adults (no children)	17.3	26.1	27.8	31.1
2 or more adults + children	21.8	19.8	21.4	18.1
3 or more adults (no children)	9.2	8.2	12.3	11.7

Figure 6.4.6 Households in the priority areas

Households	Wester Hailes	Craigmillar/Niddrie	North Edinburgh	City
Total households	377	656	352	4124
	%	%	%	%
Single people	50.9	62.4	55.7	48.0
Single parents	27.3	19.2	27.0	28.0
2 adults (no children)	8.2	9.2	8.2	9.1
2 or more adults + children	10.1	5.3	4.3	9.2
Single + others over 16 years	1.9	3.4	4.5	3.2
Couple + others over 16 years	1.6	0.6	0.3	1.5

Figure 6.4.7 Housing allocations in the priority areas and city wide

	Partial Housing Benefit %	Full Housing Benefit %	Total %
Wester Hailes	33	67	76
Craigmillar/Niddrie	28	72	81
North Edinburgh	36	64	78
City	42	58	73

Figure 6.4.8 Housing benefits in priority and city wide

	Priority Areas %	City-wide %
Council houses	71	19
Owner occupied	19	67
Council house sales (since 1980) nos.	2000	17 000

Figure 6.4.9 Housing in priority areas and city wide

Apt Size (rooms)	Wester Hailes %	Craigmillar/Niddrie %	North Edinburgh %	City %
1	0	1.0	0.2	1.2
2	30.7	17.5	14.5	21.8
3	45.2	62.5	62.0	54.9
4	20.0	17.0	21.0	19.0
5+	4.1	1.8	2.5	2.8
Total houses	4329	3821	6271	37 013

Figure 6.4.10 Apartment size in priority areas and city wide

	%
Wester Hailes	19.5
Craigmillar/Niddrie	18.5
North Edinburgh	17.0
City	9.3

Figure 6.4.11 *Unemployment rates in priority areas and city wide*

STUDENT   **6.10**   ACTIVITY

1 Study all of the above statistics, and then write an explanatory account of the reasons why these three areas were given priority status for housing.
2 In your opinion, which of the three areas most justifies priority status?

STRATEGY FOR THE PRIORITY AREAS

Since 1988 Edinburgh City Council has pursued a strategy aimed at concentrating limited resources on the priority areas in the city. The strategy recognises the fact that available resources are insufficient to achieve a successful resolution of the problems, and therefore additional resources will be needed from elsewhere, including the private sector.
Four specific objectives have been identified:
● To arrest the population decline
● To improve the houses and the environment
● To provide housing to meet current and projected needs
● To link housing investment to local training and employment initiatives.
In order to achieve these objectives, Edinburgh City Council aim to:
● reduce the number of surplus houses in the three areas;
● encourage tenure change in order to improve choice, bringing the areas more into line with the city-wide pattern;
● transfer both land and houses to local housing associations and to private developers for low-cost home ownership.
The main aim is to increase owner-occupation in Wester Hailes to a level of approximately 25 per cent, and to approximately 16 per cent in

Craigmillar/Niddrie. Levels of owner-occupation in North Edinburgh, at present higher than in the other two areas, can be sustained by 'right-to-buy' sales. House-building for rent will continue in all three areas, but the main agencies will be housing associations and housing co-operatives rather than the City Council.

STUDENT   **6.11**   ACTIVITY

1 As a young housing executive, outline the arguments that you would deploy to encourage tenants to move from council housing into home ownership.
2 Explain why you would get different reactions from different age groups.
3 Using the profiles of the three priority areas as a guide, in which area do you think you would have most success?

The Wester Hailes Partnership: the broader scene

In 1988, the Scottish Office published a document, 'New Life for Urban Scotland'. It proposed the creation of four partnership programmes in locations in four different Scottish cities: Glasgow, Paisley, Dundee and Edinburgh (at Wester Hailes, one of the priority areas for housing in the city). Three main principles for the partnership programmes were established by the Scottish Office:
● an integrated approach to economic, social and physical regeneration;
● the inclusion of the private sector in partnership to bring expertise, advice and resources;
● the full involvement of the local community in the decision-making process, thus encouraging people to take responsibility for their own area.
By any standards Wester Hailes was a deprived area, although Edinburgh City council initially resisted the idea of partnership status because it was felt that Craigmillar had a greater claim. Although Wester Hailes was desperately short of social and welfare facilities in the initial years of people moving into the estate, concerted community action over the next five years led to a number of important initiatives. The existence of strong community activities in Wester Hailes was quoted as a prime reason for its selection for the partnership programme.

PARTNERSHIP AIMS IN WESTER HAILES

The 1994 Partnership Progress Report quoted seven main aims for the partnership:

- Housing: to create demand for housing in Wester Hailes across the full spectrum of types of house and types of tenure;
- Environment: to create a mix of high quality facilities and environment of sufficiently high quality that at least 50 per cent of people questioned in any market survey state that they were 'quite satisfied' or 'very satisfied' with the facilities and the general environment in Wester Hailes;
- Publicity: to enhance the image of Wester Hailes in the eyes of residents, investors, employers and the wider community;
- Employment and Training: to halve the gap between the Wester Hailes unemployment rate and the substantially lower one for Edinburgh over five years;

- Economic Development: to develop a community that has the appropriate level of economic activity and services for its size and which also has taken advantage of its location on the edge of a rapidly growing economic development area;
- Social Policy: to increase average household income in real terms;
- And, more generally, to give the community a stake in the new assets to be created locally and a share in the process of their creation.

1 Draw a flow diagram to show how all of these aims are inter-related in a programme of community action
2 Use a technique to rank these aims in order of importance
3 For the first three of the ranked aims, suggest a programme of future action which would help these aims to be realised.

You will need to refer to the profile of Wester Hailes at the beginning of this section

6.5 Managing A New Town: Craigavon, Northern Ireland

It was seen that, in the Dublin Region, the building of three new towns to the west of the city was a response to increasing population pressures in the zone beyond the city limits. Craigavon, Northern Ireland's only new town was built for rather different reasons. In part, it was designed as an overspill to relieve urban pressures on Belfast (some 50 km away), but it also had the purpose of stimulating growth and development in the south west of Ulster. Incorporating the two existing towns of Lurgan (Figure 6.5.1) and Portadown (Figure 6.5.2), Craigavon was to have a linear pattern with the two existing towns at either end of an urban corridor that would include the entirely new settlement of Brownlow in the centre (see Figure 6.5.3). It will be seen from the map that Craigavon Borough takes in a

considerable rural hinterland, and has an extensive frontage along the southern shores of Lough Neagh. A planned population of 120 000 was forecast for the 1980s, rising to 180 000 by the end of the century, although these optimistic figures

Figure 6.5.1 *Lurgan, Northern Ireland*

Figure 6.5.2 *Portadown, Northern Ireland*

were later substantially revised downwards. Craigavon never really developed as was intended in the early 1960s. Several factors were responsible for a much slower growth than was originally envisaged:

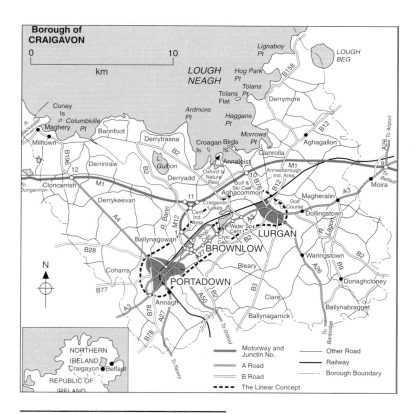

Figure 6.5.3 *Craigavon, general structure*

• economic recession meant that inward investment targets for industry were not met;
• the policy of population dispersal from major cities to overspill centres was reversed;
• the 'Troubles', perhaps unforeseen in the early 1960s acted as a major deterrent to investment and expansion of business.

The need for a new town in the region had virtually disappeared by the 1980s, and Brownlow (Figure 6.5.4), the new settlement between Lurgan and Portadown was left as a series of 23 housing estates, planned for a high level of car ownership, located in what was still in essence a rural area. Social, welfare and leisure facilities that could be expected in such an area were never built. An industrial estate and regional shopping centre do exist, but they are some distance from the main housing areas. However, much of the housing stock in Brownlow is of a reasonable quality, and is low density. New town planning gave Brownlow a good road infrastructure, although for those without cars the bus service is relatively poor. A measure of the changing status of Craigavon is that its population, planned for 180 000 by the end of the century was, in 1995, a mere 52 700 with Brownlow (the new element) having just over 9000 inhabitants. The projected population for 2010 is now estimated at 59 000.

Management of a new town whose status has clearly been down-graded faces a series of problems in the late 1990s. Craigavon, as a linear new town has four separate foci, Portadown, Lurgan, Brownlow and the Craigavon Central Area (which was to be the administrative and commercial core of the New Town). Two of these foci are traditional market towns, Brownlow is principally a series of residential estates, and the Central Area is still something of a separate entity. Craigavon as a whole has not benefitted from any recent urban renewal programme. Long term unemployment has been the norm in many housing estates, and part of Brownlow has, until recently, qualified for EU assistance as an area of acute poverty.

The Craigavon Area Plan, the essential management tool, seeks to put these problems into focus

1 Although the Craigavon Urban Area Plan envisaged that most new

Figure 6.5.4   *Brownlow and Central Area, Craigavon*

development would take place between Portadown and Lurgan, consonant with the idea of a linear new town, the Area Plan proposes that new developments should occur in both Portadown and Lurgan, whilst Brownlow should undergo a period of consolidation.

2 Both Portadown and Lurgan are allocated new areas for housing, whilst in Brownlow, regeneration of housing estates is the main objective, building a number of important community-based efforts.

3 Portadown is the primate shopping and service centre in Craigavon Borough, and its status will be maintained and improved. Although Lurgan serves a smaller catchment, its character as a traditional market town will be retained. In Brownlow basic shopping facilities at neighbourhood level need improvement, whilst the Central Area will be developed as a focus, not only for the linear area, but for a wider area by virtue of its excellent road infrastructure.

STUDENT   ACTIVITY

1 To what extent do the new proposals in the Area Plan meet the needs of changed circumstances in the Craigavon area?

2 As a prospective industrialist, wishing to take advantage of one of the vacant industrial sites in the Borough, what reassurances would you need concerning the quality of life of your local workforce, who would live in Brownlow, Portadown and Lurgan?

ESSAYS

1 The underprivileged are no longer confined to the inner city. Discuss the validity of this statement.

2 Urban transport is likely to emerge as the key issue in cities in the early twenty first century. To what extent do you agree with this statement?

FIELDWORK OPPORTUNITIES & PROJECT SUGGESTIONS

1 Select an area of a city where urban renewal is taking place (this could be a City Challenge Area or a Single Regeneration Budget area, or part of either). Carry out a housing quality survey for selected parts of your area to show contrasts. Draw maps to show the level of take-up of improvement grants in different streets. Carry out an environmental quality survey in different streets to show contrasts. Map improvements in environmental quality, e.g. pedestrianisation, traffic calming, provision of new street furniture etc. Assess the success of the renewal scheme using a numerical scale from different criteria.

2 Compare a run-down inner city shopping area, with a new urban fringe shopping centre. Assess shopping quality in the two different areas. Institute an environmental quality (EQ) survey of the two different areas (you could use a grid for establishing EQ values over the two different areas). Assess the importance of the private car to the shoppers in the two different areas (look at car parking capacity, and estimate levels of car traffic entering and leaving car parking facilities in the two areas.

7 Managing the Industrial Environment

KEY IDEAS

- De-industrialisation affects all of the traditional industrial areas of Britain
- Employment in manufacturing industry is being replaced by employment in the service industries
- Management in industrial areas seeks to create a healthy economic climate which enables local businesses to grow, and encourages inward investment
- Decisions in industrial location and investment are increasingly taken by transnational companies operating in a global context
- Environmental considerations, such as the control of air and water pollution, are now an integral part of the management of the industrial environment

7.1 Introduction

The comment by the taxi driver effectively sums up some of the principal issues involved in the management of the industrial environment in the 1990s. However, it is important here to make a distinction between Britain and the Republic of Ireland, since the forces of change operating in the industrial environment are rather different in the two areas. While there has been a steady decline in the number of manufacturing jobs in the United Kingdom, largely through **de-industrialisation**, until recently there has been a steady increase in jobs in manufacturing industry in the Republic of Ireland.

In the United Kingdom, management of the industrial environment in the second half of the twentieth century has been much concerned with both the organisation of industry, and the changes consequent upon a declining manufacturing base.

Three major trends in the organisation of industry may be noted:
- privatisation: the policy of transferring industry from the public to the private sector has been a major feature of the Conservative Governments in the 1980s and early 1990s. It has privatised industry in the primary producing sector, e.g. coal, the manufacturing sector, e.g. steel, the former public utilities, e.g. gas, electricity and water, and in the service sector, e.g. telecommunications. One important consequence of privatisation has been the rationalisation of these industries with the consequent loss of jobs;
- the increasing importance of **transnational** companies. Both manufacturing industry, and to a lesser extent, service industries are more and more dominated by these companies;
- inward investment by foreign companies is becoming increasingly important in

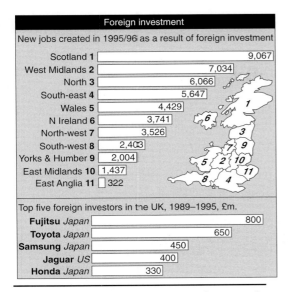

Foreign investment

New jobs created in 1995/96 as a result of foreign investment

Scotland 1	9,067
West Midlands 2	7,034
North 3	6,066
South-east 4	5,647
Wales 5	4,429
N Ireland 6	3,741
North-west 7	3,526
South-west 8	2,403
Yorks & Humber 9	2,004
East Midlands 10	1,437
East Anglia 11	322

Top five foreign investors in the UK, 1989–1995, £m.

Investor	£m
Fujitsu *Japan*	800
Toyota *Japan*	650
Samsung *Japan*	450
Jaguar *US*	400
Honda *Japan*	330

Figure 7.1.1   *What foreign investment has meant for the UK*

the United Kingdom, and has been particularly important in stimulating job growth in Scotland, the West Midlands and the North. Figure 7.1.1 shows the number of jobs created through foreign investment in the various regions of the United Kingdom in 1995–6. It also shows the top five foreign investors in the United Kingdom in the period 1989–95.

De-industrialisation in the British economy, and the consequent loss of jobs (particularly serious in the recession of 1979–81) has prompted a number of initiatives to promote industrial regeneration in the regions most badly hit:

● Government regional policy: the creation of Assisted Areas had its origins in the 1930s, but has persisted as a tool to redress regional imbalance and stimulate recovery in regions of high unemployment. The latest designation of Assisted Areas in the United Kingdom in 1993 is shown in Figure 7.1.2. Significant additions to the map were those in the east and south east of England;

● the creation of regional agencies (with the status of **quangos**): these were designed to promote economic growth in some regions, e.g. Welsh Development Agency, and the Scottish Development Agency (later renamed Scottish Enterprise);

● the development of Enterprise Zones: this was a Government initiative aimed at stimulating business in decaying urban and industrial areas. The first zones were designated in 1980, and the last in 1994, and, with a life of ten years, their influence will persist into the next century;

● Urban Development Corporations: although these were created in the 1980s to aid regeneration in run-down urban areas, often in the inner city, they have had an important effect on the management of the industrial environment (see Figure 7.1.3);

● Training and Enterprise Councils: these were established in England and Wales in 1989 to involve private enterprise in local economies.

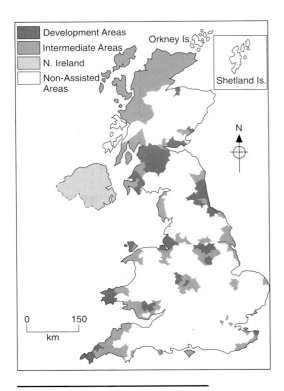

Figure 7.1.2   *UK Assisted areas, 1993*

Figure 7.1.3   *Redeveloping Liverpool's Docklands, Merseyside Development Corporation*

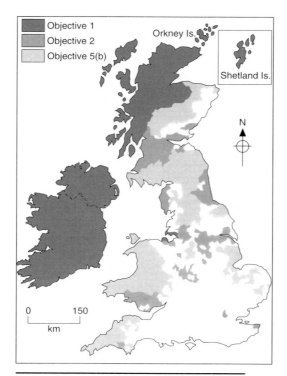

Figure 7.1.4 *UK and Republic of Ireland areas qualifying for aid from the EU's reformed structural funds, 1994*

as peripheral. Three objectives are identified for the purpose of European Aid:

Objective 1: regions whose development is lagging behind, i.e. regions whose per capita **GDP** is at or below the level of the Community average;

Objective 2: the economic conversion of declining industrial areas;

Objective 5b: economic diversification of rural areas.

Figure 7.1.4 shows the areas of the United Kingdom and the Republic of Ireland that qualify for European Aid under the different Objectives.

Manufacturing industry in the Republic of Ireland has seen a steady growth in both absolute numbers and in percentage terms in the second half of the twentieth century, although there has been some decrease in the late 1980s. Although Government influence over industry in the Republic was originally protectionist, after 1959 it embarked on a programme designed to attract inward investment from outside. Although the Buchanan Report (see page 113) encouraged the idea of growth centres, the Five Year Plans of the 1970s opted for a disproportionate degree of growth in the marginal areas. As in the United Kingdom, the second half of the twentieth century has seen a considerable growth in foreign investment in the Republic, particularly from the United States. Aid from the European Regional Development fund is particularly important, since the whole of the Republic has an Objective One status (see Figure 7.1.4).

Of increasing significance has been the involvement of the European Union in support of areas in both the United Kingdom and the Republic of Ireland that have lagged behind in economic growth. In as far as the core of the EU's economic activity is seen in a zone from the English Midlands to Northern Italy (the 'hot banana'), much of the United Kingdom and all of the Republic of Ireland are seen

7.2 Case Studies in the Management of the Industrial Environment

Four case studies are used in this Chapter, to explain how industrial change is managed. The first study is of Northern Ireland, with special reference to Belfast. Traditional industries such as heavy engineering (shipbuilding) and textiles (linen) gave a distinctive character to the economy of Belfast and other towns in the east of the Province. Elsewhere the industrial base was poorly developed. In

the second half of the twentieth century, Northern Ireland remained the poorest region in the United Kingdom in terms of standards of living, and was beset by levels of unemployment consistently higher than the United Kingdom average. Managing change in Northern Ireland is concerned with broadening the industrial base, and attracting inward investment.

Blackburn is a traditional textile town in

the north west of England and, like many other towns in the area has witnessed the long-term decline of the traditional industrial bases of textiles (cotton) and engineering. Attracting new industry is only one part of management activity, since the regeneration of the area's infrastructure is essential to the future growth of the local economy. Peterborough provides a marked contrast to the other centres studied. Although it did have some industrial base, it has managed industrial change by capitalising

on its location between East Anglia and the English Midlands, and attracting a range of new industries, both in the manufacturing and service sectors. To complete the case studies, a brief review of the growing industrial attraction of the Irish Republic is made. Although Dublin remains the most important single industrial centre in the Republic, the west of Ireland has proved to be an attractive area for inward investment in the second half of the twentieth century.

7.3 Managing Industrial Regeneration in Northern Ireland

STUDENT **7.1** ACTIVITY

1 Study the extract from the article 'Major announces jobs programme for Ulster' (Figure 7.3.1)
  a Why is new foreign investment so essential for the regeneration of Ulster's economy?
  b Suggest what some of the disincentives to foreign investment in Northern Ireland may have been?

  c What encouraging features of Northern Ireland for inward investment are noted in the extract?

The map (Figure 7.3.2) shows unemployment rates by travel-to-work area (**TTWA**) in Northern Ireland in 1994. There are clear variations within the Province, with the highest levels being in the more peripheral areas in the west and in the south-east around Newry. Only the

Major announces jobs programme for Ulster

JOHN MAJOR yesterday announced a new programme to tackle long-term unemployment as part of efforts to rebuild the Northern Ireland economy in the wake of the IRA and loyalist cease-fires.

A larger and potentially more important trade and investment conference is being planned by the Clinton administration to take place in Philadelphia in April. Yesterday Ron Brown, US Commerce Secretary, said he would present that conference with a plan for speeding up growth in the Northern Ireland economy.

Mr Major commended the quality of the local workforce, its low rate of labour disputes and the incentive packages available. He also announced five examples of new investment decisions by existing companies, one creating 100 jobs at a new business park in west Belfast. He announced a pilot scheme aimed at offering the long-term unemployed work for up to three years. It will have 1,000 places for those who have been out of work for more than a year.

The Independent
15 December 1994

Figure 7.3.1 Major announces jobs programme for Ulster

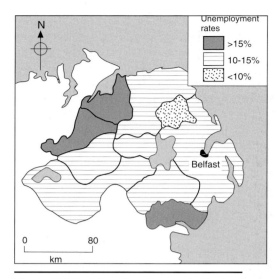

Figure 7.3.2 *Unemployment rates in Northern Ireland, August 1995*

- its peripherality: it lies on the extreme north-western edge of EU countries, and this clearly creates transport problems;
- a high rate of natural increase in the labour force (many entering the potential labour force find it extremely difficult to obtain a job);
- a small manufacturing base, which exhibits a low level of productivity in the indigenous sector (as distinct from the sector funded by inward foreign investment);
- an overdependence on the public sector (particularly in the service industries), which, conversely, has helped to cushion the Province from the worst effects of the recession of the early 1990s.

Managing the regeneration of the economy in Northern Ireland is the responsibility of the Department of Economic Development. Four main economic development agencies operate within the framework of the Department as shown in Figure 7.3.3.

The Industrial Development Board was set up in 1982, with its prime aim of promoting industrial growth in Northern Ireland. It is responsible for encouraging and assisting the development of indigenous industry, and also has the challenging task of attracting inward investment from foreign-based companies. In the year 1994–5, the Industrial Development Board negotiated 76 projects with indigenous and foreign companies which had the potential to create 4000 new jobs. Of these 31 were companies with headquarters outside of Northern Ireland, with plans to create 3200 new jobs through

Ballymena TTWA compares favourably with the average rate for the United Kingdom. The map does conceal variations in the rate of unemployment within the TTWAs. For instance, some areas in West Belfast have unemployment rates of greater than 50 per cent. One of Northern Ireland's biggest problems is the large proportion of long term unemployed (more than one year): more than half of those unemployed were in this category compared to approximately one-third in the United Kingdom as a whole.

Apart from the high levels of unemployment, which has long been one of the most intractable problems in Northern Ireland, the Province has had to cope with other inherent difficulties:

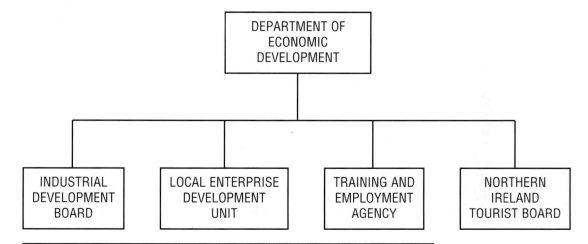

Figure 7.3.3 *Organisation of Department of Economic Development, Northern Ireland*

expansion and greenfield investment projects worth approximately £226 million. Attracting inward investment to Northern Ireland depends on a number of important incentives:

- financial incentives: amongst those available are cash grants of up to 50 per cent for buildings and machinery, and additional grants towards rents, company development, marketing strategy, and research and development;
- a range of suitable greenfield sites, strategically sited within Northern Ireland;
- an on-going support service in the early years after start-up;
- relatively low labour costs compared to other international competitors.

The success achieved in attracting foreign inward investment is indicated on the map showing the number of firms locating in Northern Ireland (Figure 7.3.4).

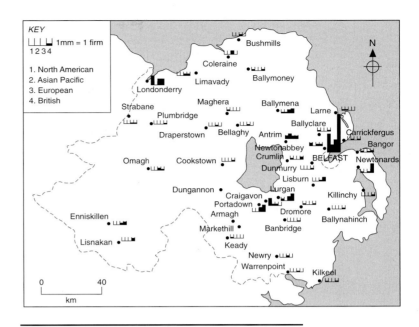

Figure 7.3.4 *Number of firms locating in Northern Ireland*

STUDENT **7.2** ACTIVITY

1 Examine Figure 7.3.4 and comment on the spatial distribution of new investment in Northern Ireland.
2 Refer back to the map showing unemployment rates in Northern Ireland. How successful has the Industrial Development Board been in attracting new industry into areas of high unemployment? How does this compare to your answer to question 1?
3 Study the comments from some foreign business executives, referring to the reasons why they chose to locate in Northern Ireland. Construct a table to show the factors they mention.

'AVX has had a presence in Northern Ireland since 1979 and has found the location to be ideal, allowing us to service our customers on a European and world-wide basis in a highly competitive market.'
AVX (Kyocera) – electronic components (Japan)

'Global companies like Du Pont operate in a highly competitive environment. Our local workforce has proven it is as good as – if not better than – others in the world.'
Du Pont synthetic fibres and rubber (USA)

'With our new Londonderry yarn-spinning facility together with the fast and cost-effective transportation services available here, we have reduced delivery times from anything up to nine weeks to around one week.'
Fruit of the Loom – cotton yarn (USA)

'We are so satisfied with the local transportation infrastructure that we have never considered our Northern Ireland Plant to be at any disadvantage.'
Optibelt – power transmission belts (Germany)

'We can easily supply engine plants in Cologne and Valencia on the Continent and Dagenham and Bridgend in Great Britain on a **Just-in-Time** basis.'
Ford – automotive parts (USA)

Strategy for the late 1990s sees the Industrial Development Board committed to creating an industrial environment that will compete very effectively in a global economy (1998 targets shown):

- to encourage indigenous companies to broaden their export markets (annual increase of 7 per cent in value of products and services sold outside Northern Ireland);

- to continue to attract inward investment from foreign companies, particularly in the **sunrise industries** (60 inward investment projects with a potential of 12 000 jobs);
- to expand the operations of existing foreign companies that are already operating successfully in Northern Ireland;
- to promote the development of inward investment in the most disadvantaged areas in Northern Ireland (up to 75 per cent of the new inward investment projects to be in the disadvantaged areas).

STUDENT ACTIVITY

1 Using material from all of this section on industrial regeneration in Northern Ireland, discuss
   a the role of transnational companies
   b the participation of the Province in the global economy
   c the role of Government intervention to enhance the environment in which industry has to operate.
2 What arguments could be deployed to attempt to persuade a foreign company to invest in a plant in Northern Ireland rather than in South Wales or Central Scotland? (Both have excellent records of attracting such investment.)

CREATING A STIMULATING INDUSTRIAL ENVIRONMENT IN BELFAST

To some, Belfast's role as an industrial centre is a surprising one. In the 1950s two writers commented that Belfast 'succeeded in becoming an important centre of manufacture even though it is practically devoid of native sources of minerals and power, is separated from the principal domestic markets by the Irish Sea and for most goods provides only a very small market of its own.' Notwithstanding these disadvantages, Belfast flourished as a major industrial city within the United Kingdom in both the nineteenth and early twentieth centuries, its industrial roots firmly fixed in textiles, ship-building and other forms of engineering. Since the middle of the twentieth century, Belfast, in common with so many other British cities, has suffered from de-industrialisation. In the years between 1926 and 1981 employment in textiles fell from 51 000 to just over 1000; shipbuilding saw a contraction from 19 000 jobs to just under 12 000 (Figure 7.3.5). Since 1981 further job-shedding has occurred, so that in 1993 the total number employed in manufacturing industry in Belfast was only just over 19 000.

STUDENT ACTIVITY

Study the extract from the *Independent*, Figure 7.3.6.
   a Identify the evidence in the extract which shows that Shorts are linked into a complex global network in the aerospace industry.
   b How does the situation described in the article fit into the pattern of de-industrialisation in Belfast?

The diagram, (Figure 7.3.7), shows that, as jobs in manufacturing industry have declined, so jobs in the service industries have grown, to the extent that, in 1971 service industries provided 60 per cent of all employment in Belfast (already at that time a clear symptom of continuing de-industrialisation), but by 1993 this proportion had risen to 84 per cent. The map, Figure 7.3.2, shows that the Belfast TTWA had one of the lower levels

Figure 7.3.5   Harland and Wolff Shipyards, Belfast

Fokker fall threatens 1,000 jobs at Shorts

THE collapse yesterday of Fokker, the Dutch aircraft manufacturer, put just over 1,000 jobs at risk at Shorts, the Belfast aerospace company

The bankruptcy of Fokker led to the immediate lay off of 660 workers at Shorts employed building wings for the failed company's aircraft. The future of several hundred other ancillary jobs is also threatened.

Shorts said it had already cut the number of jobs at risk from the 1,460 who were put on protective redundancy notice in January, when a Dutch government backed attempt to save Fokker began.

Shorts added that it hoped to reduce the number of redundancies among its workforce of 6,800 to below 1,000 by redeployment and training schemes until new orders came in.

If is feared hundreds of jobs in supply companies could be hit, while City sources said the bankruptcy could cost Rolls-Royce, which makes engines for Fokker, up to £30m. Rolls said it was too early to say whether jobs would be lost.

Shorts has been told that some Fokker 100/700 aircraft are to be completed over the next three months, which will provide some work.

A spokesman for Shorts, which is owned by Bombardier, a Canadian engineering and aerospace group, said: "We are now vigorously exploring a range of measures to limit the effect of Fokker's bankruptcy. In addition we are continuing to pursue new business opportunities which include several UK government defence programmes for which we are currently bidding."

Shorts has insisted throughout the Fokker crisis that its future is safe because Bombardier has injected £200m since 1989 to pay for diversification and new technology.

The failure of two Far Eastern buyers to come up with offers to buy Fokker finally killed off the company after desperate late night negotiations on Thursday with Samsung of South Korea. "This means the end of 77 years of aircraft history in the Netherlands," said Ben van Schaik, Fokker's chairman.

The company was put in the hands of administrators in January when its controlling shareholder, Daimler-Benz of Germany, withdrew support because of mounting losses. But Fokker was given Dutch government bridging finance to keep it alive as talks with potential buyers continued.

At Fokker, more than 5,600 workers involved in aircraft manufacture will be dismissed, the largest single redundancy in Dutch history. But 960 will be offered jobs at the remaining divisions that escaped collapse. A number of viable businesses employing 2,500 staff in aircraft maintenance, electronics systems and special products will be lumped into a company called Fokker Aviation. However, the rump company will still need new backing, Fokker said.

Samsung Aerospace said: "We were interested in Fokker as a strategic tie-up . . . But we were not able to make the final offer due to time constraints."

Hans Wijers, Dutch economics minister, said of the talks: "The only thing we got was a letter which contained less commitment than earlier signals." A second suitor, China Aviation Industries, had earlier decided against an offer.

Since it was founded in 1919, Fokker has built more than 125 different types of aircraft.

The Independent 16 March 1996

Figure 7.3.6   *Fokker fall threatens 1000 jobs at Shorts*

of unemployment in Northern Ireland, at 11.1 per cent. Yet it was still higher than the United Kingdom average, and conceals much higher, and much more worrying levels of unemployment within the city.

Within the City Council area, there is an unemployment level of 25 per cent. Highest rates were in the inner city, and in the west of the city, and the worst areas there, it has been noted, had more than 50

Year	1961	1966	1971	1975	1981	1985	1991	1993
Manufacturing and Construction (%)	48.3	45.9	39.9	36.9	29.5	25.0	19.1	16.7
Services (%)	51.7	54.1	60.1	63.1	70.5	75.0	80.9	83.3
Percentage of Employees in Manufacturing and Construction and in Services.								

Figure 7.3.7   *Belfast urban area 1961–93, Employees in manufacturing and construction services*

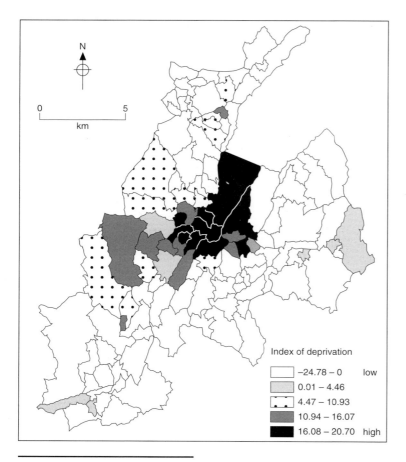

Figure 7.3.8 *Deprivation in Belfast*

Index of deprivation

☐	−24.78 – 0	low
☐	0.01 – 4.46	
⋮	4.47 – 10.93	
▨	10.94 – 16.07	
■	16.08 – 20.70	high

per cent of their workforce unemployed. In these areas older males have often borne the brunt of de-industrialisation losses, and unemployment is also high amongst those under 30, many of whom are coming on to the labour market for the first time. In both north and west Belfast 60 per cent of all the unemployed were in the long-term category. The map (Figure 7.3.8), shows levels of deprivation in the Belfast urban area, and confirms the low quality of life experienced in the inner city and in parts of west Belfast.

Two important initiatives to stimulate industrial recovery were introduced in the 1980s:

- the creation of the Belfast Enterprise Zone (see Figure 7.3.9) in 1981. Although it has now expired (all Enterprise Zones have a life of ten years) it did have some success in creating jobs. However nearly 70 per cent of the jobs were in the service sector, and the Harbour section of the Zone fared significantly better than the Inner City area.

- the Making Belfast Work project; this aims to target the most deprived areas within the city (see Figure 7.3.8) in a concerted effort, involving both public and private sector and local communities, to improve their social, environmental and economic status. Its aims for the future are to be channelled through a City Regeneration Board, to be established in July 1996.

Figure 7.3.10 *Laganside, Belfast*

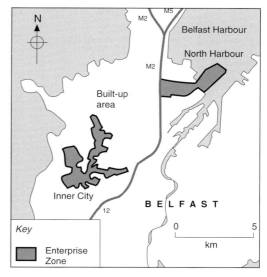

Figure 7.3.9 *The Belfast Enterprise Zone*

THE BELFAST ECONOMY: ACTION FOR THE FUTURE

In 1994 Belfast City Council produced their Economic Development Strategy, which consisted of an analysis of the current economic status of the city, together with a strategy for the future. Part of the investigation of the City's current economic status was a SWOT analysis (Strengths, Weaknesses, Opportunities and Threats). This is given, in abbreviated form, below.

SWOT Analysis

Strengths

1 Belfast is the centre of economic activity in Northern Ireland. It has 18 per cent of the Province's population, but 30 per cent of Northern Ireland's manufacturing companies and 21 per cent of all retail activity.
2 Belfast has received massive infrastructure investment in the last decade – up to £800 million in housing, offices and retail facilities.
3 It has a well-integrated internal and external transport infrastructure.
4 It has a sound telecommunications network, amongst the most sophisticated in Europe.
5 Infrastructure development over the last five years has stemmed in part from the activities of public sector agencies in Belfast, which have had to meet specific development targets, set by government departments.
6 Belfast has a justified reputation of high standards in secondary, further and higher education.
7 The Council has purchased sites for development, such as the Gasworks site and has developed a partnership with the Laganside Corporation (see Figure 7.3.10).

Weaknesses

1 The image of Belfast is one of perceived violence and deprivation.
2 This image has tended to discourage inward investment, and has reduced motivation amongst the business community.
3 The peripherality of Belfast within the EU has discouraged foreign investment

from non-EU countries seeking a foothold in the Union.
4 There is an over-reliance on public sector jobs, partly a reflection of de-industrialisation, and partly a result of the centralisation of public sector agencies in the city.
5 The relatively high levels of unemployment in Belfast, and its position relative to the rest of the United Kingdom in this respect.
6 Significant pockets of poor educational achievement at primary and secondary level.

Opportunities

1 The cessation of violence, consequent on the ceasefire declared by loyalist and republican paramilitary groups, although since 1996 and the resumption of violence both on the British mainland and in Northern Ireland this is, at the time of writing, a doubtful opportunity.
2 Changing political environment in the late 1990s may bring additional European funding, together with promises of aid from the United States, and possible National Lottery funds from the United Kingdom.
3 Well-established development support organisations should ensure further development of infrastructure.
4 Opportunities to project a more positive image of Belfast.

Threats

1 The entrance of considerable numbers of adolescents into the labour market, thus increasing the pressure on jobs.
2 Peripherality means difficulty competing within Europe for investment.
3 Changing development priorities mean that Northern Ireland can no longer depend on grant aid to support the economy.
4 The continuing negative image that has discouraged inward investment, and forced Belfast to depend on European aid.

1 Use the SWOT analysis, together with other material from this section, to produce a list of the most important strategic economic development issues facing Belfast over the next five years.
2 Develop an outline action plan, the principal aim of which is to create a stimulating industrial environment in the Belfast area.

7.4 Coping with Change in Blackburn

Although at the present day one of the most important of the many industrial centres which assist in retaining for Lancashire its world wide fame, the town of Blackburn cannot lay claim to much historical or traditional antiquity.

Introduction to a directory of industry in Blackburn, published in 1889.

In 1889 Blackburn was indeed one of the leading industrial centres in Lancashire, and the directory entry for Thomas Dugdale, Brother & Co., Spinners and Manufacturers, Griffin Mills, Livesey, Blackburn (Figure 7.4.1) was typical of the cotton mills that dominated the industries of Blackburn at that time. It had also acquired an important reputation for the production of textile machinery, and James Haydock's mill (Figure 7.4.2) also appeared in the directory.

Just over a hundred years later, Blackburn Borough Council produced a document 'The Changing Face of Blackburn and Darwen', which reports on a very different state of industrial affairs. Figure 7.4.3, taken from the document shows the relative proportions of employees in manufacturing industry and in the service industries in Blackburn, the North West and in the United Kingdom. Although Blackburn, like most towns in the United Kingdom now, is dominated by the service industries, it is still very much an industrial town, with 10 per cent more employees in manufacturing industry than in the North West and 12 per cent more than in the United Kingdom as a whole. The town has had to face de-industrialisation on a massive scale. For instance, in 1929, 62 per cent of those employed in Blackburn were in the textile industry, yet by 1993 this had fallen to a mere 6 per cent. There has been

Thomas Dugdale, Brother & Co., Spinners and Manufacturers, Griffin Mills, Livesey, Blackburn.—This great business was founded in 1852, and its entire history has been a record of notable commercial ends accomplished and high prosperity achieved by the energetic enterprise and sound principles of its proprietors. In 1874 the senior partner, Mr Thomas Dugdale, died, and since then the house has been under the control of the brother, Mr Adam Dugdale, whose administration admirably illustrates the qualities of activity, experience and judgment which mark his personal character. The firm have their headquarters at the well-known Griffin Mills, and the industry there carried on exhibits many features of splendid vitality and thorough development. The mills are two in number, and together cover an area of 13,000 square yards. Their mechanical equipment and general facilities are of the most elaborate and well-considered description, and the fine plant in operation includes 1,768 looms and 99,000 spindles, by machine makers of first-class repute. Motive power for this immense outfit of machinery is supplied by ten full-size Lancashire boilers (30 feet by 7 feet) and two hot-water boilers (100 lbs pressure), driving four pairs of horizontal engines with an aggregate power of 2,500 horse. Upwards of twelve hundred hands are employed in these exceptionally extensive mills, and the scene of animation and activity presented by the entire establishment when in "full swing" (as it nearly always is) affords an excellent clue to the magnitude of the trade that necessitates such a vast expenditure of productive energy. Messrs. Dugdale use American cotton chiefly, though a certain quantity of Egyptian cotton is also employed in their industry. The spindle products of the Griffin Mills embrace a very large output of "Medium Counts" in yarns; and the loom products comprise principally plain shirtings for the India and China markets. The trade controlled is one of very widespread range and extent, and the house is certainly among the most influential and typical in this important centre of the cotton industry. Mr Adam Dugdale, J.P., the head of the firm, is an Alderman of Blackburn, and has twice held office as Mayor of the borough. Ever since he first became associated in person with the manufacturing undertakings of the town he has been highly esteemed for his active and earnest participation in all local affairs. In the discharge of his official functions he has displayed an ability akin to that put forth by him in the management of his business; and the excellent institution of the Church of England schools at Blackburn, founded and built by the late Mr Thomas Dugdale, and now supported entirely by Mr Adam Dugdale, the present owner of these mills, stands as a permanent remembrancer of the deep and beneficial interest this respected gentleman has always taken in the improvement of educational resources and facilities throughout his neighbourhood.

Figure 7.4.1 Thomas Dugdale & Co entry in Blackburn Trade Directory, 1889

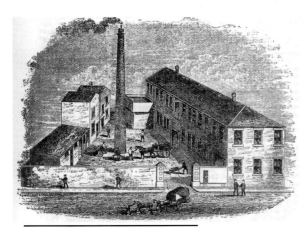

Figure 7.4.2 *James Haydock's mill*

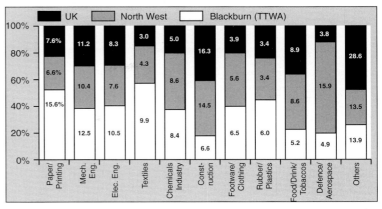

Figure 7.4.4 *Comparative industrial sector breakdown, Blackburn, N W England and United Kingdom*

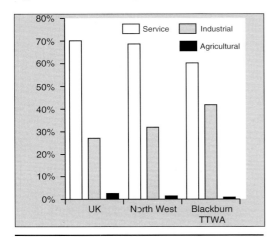

Figure 7.4.3 *Comparative size of industrial sectors, Blackburn, N W England and the UK*

a similar decline in other industries such as textile machinery manufacture.

Figure 7.4.4 shows a comparison between the different sectors of manufacturing industry in Blackburn, North West England, and the United Kingdom.

STUDENT **7.5** ACTIVITY

1 Comment on the difference between the sectors of manufacturing industry in Blackburn and the North West of England.
2 How much do they differ from the national pattern?

STUDENT **7.6** ACTIVITY

Use all of the statistics on employment, and numbers of unemployed in Figures 7.4.5–7.4.8, to carry out a SWOT analysis of the current economic status of Blackburn.

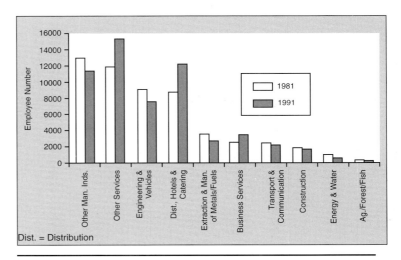

Figure 7.4.5 *Blackburn, variation of employee numbers by sector 1981–91*

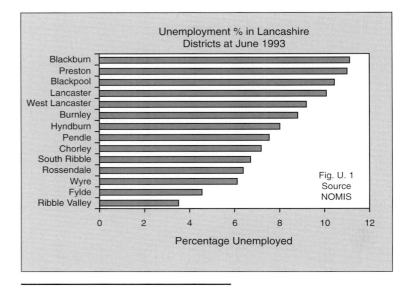

Figure 7.4.6 *Unemployment in Lancashire*

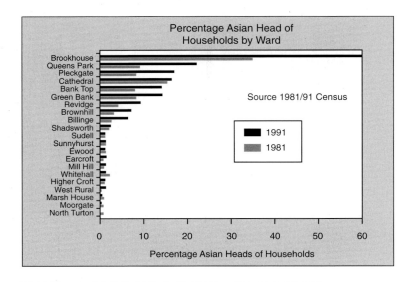

Figure 7.4.7 *Percentage of Asian heads of households by ward*

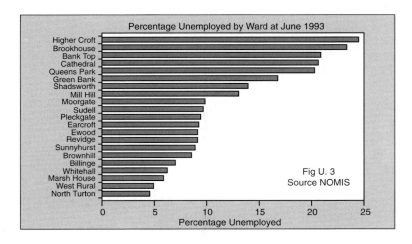

Figure 7.4.8 *Percentage unemployment by ward, June 1993*

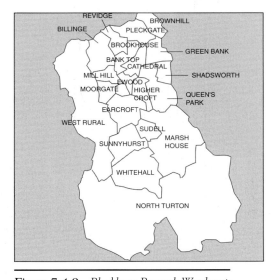

Figure 7.4.9 *Blackburn Borough Ward map*

MANAGING THE INDUSTRIAL ENVIRONMENT IN BLACKBURN

Six main issues requiring some strategic action are recognised in Blackburn Borough's own analysis of the current situation:

- registered unemployment consistently above the national average;
- a ranking within the worst third of the country in terms of its employment situation;
- long-term decline in traditional manufacturing;
- recent sectoral decline in defence and aerospace industries;
- severe unemployment in inner city wards;
- high unemployment amongst ethnic minorities.

Dealing with these issues, Blackburn has developed a raft of policies that depend on a range of funding from outside. Sources of funding include:

- Assisted Area Status: national funding
- Objective 2 Status: European funding
- City Challenge: national funding (subject to a successful bid)
- Single Regeneration Budget: national funding.

The City Challenge Scheme requires Blackburn and other authorities to compete for funding for regeneration purposes. Over a five year period City Challenge will bring £37.5 million of public sector money into the designated area within Blackburn (see Figure 7.4.10) and it is hoped to attract a further £113 million from other sources. Projects within the scheme are clearly not confined to industry alone, and include a range of housing, social and environmental initiatives. One of the most important initiatives within the City Challenge area is the Waterside programme developed along the course of the Leeds and Liverpool Canal. Along this linear artery, 13 canalside housing developments have been built, the Eanam Wharf Site has been developed to house the Blackburn Business Development Centre (Figure 7.4.11) and the Blackburn Arena, an international status ice arena has been built. Greenbank Business Park, for medium sized companies, has been developed on a canal-side site in the east of the town. Daisyfield Mill (Figure 7.4.12), one of Blackburn's most

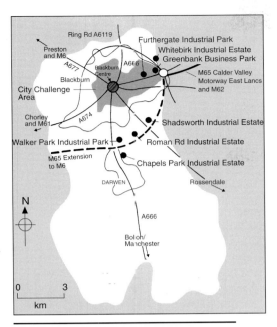

Figure 7.4.10 *City Challenge area and major industrial estates in the borough of Blackburn*

Figure 7.4.11   *Eanam Wharf, Blackburn*

prominent links with its industrial past, symbolises the transfer from manufacturing industry to the service sector. It has been completely refurbished to provide a series of office suites, and includes Granada Television's studios for east Lancashire. Money available from the City Challenge funds has been channelled into a target area, with, it has been noted, a particular emphasis on the Waterside project. In a similar way the Single Regeneration Budget (£19.5 million over seven years, with an additional £73.8 million from the private sector, and £20.1 million from other public sector investment over seven years) has concentrated much of its effort into deprived communities of Bank Top (inner city), Roman Road, Shadsworth and Darwen. Development has particularly been concentrated in the corridor that is developing along the M65 extension (see Figure 7.4.10). Several of Blackburn's industrial estates lie along this corridor, although the majority existed before the M65 extension was built.

STUDENT   **7.7**   ACTIVITY

1 Both the City Challenge and the Single Regeneration Budget funds have been targeted on specific areas within Blackburn. How effective a tool of industrial management do you consider area-concentration of funds to be?

2 To what extent will these two initiatives contribute to the long-term solution of the major strategic problems identified in Blackburn?

In an industrial town such as Blackburn, the image of the industrial environment is an important issue. The Council offers grants for Environmental Assessment (which covers such issues as types of raw materials used, methods of waste disposal, energy use, and the appearance of external zones within any industrial site), commercial improvement and industrial improvement.

Figure 7.4.12   *Daisyfield Mill, Blackburn*

7.5 Peterborough and East Anglia: Managing Growth in the Economy

Peterborough lies in the north-west corner of the East Anglia region as shown on the map (Figure 7.5.1). Its position on the border of East Anglia, and the East Midlands means that it occupies a strategic position unrivalled elsewhere in the region, and it has also enjoyed the benefits of being one of the Mark 111 New Towns designated in 1967. The city lies some 184 km from London, and it was intended that it should be an overspill centre for population from London, and become a new regional city in its own right, and a possible counter-magnet to London.

Figure 7.5.2 shows employment changes in East Anglia over the period between 1988 and 1993.
1 Summarise the main changes that have taken place in this period. Which ones do you consider to be the most significant?
2 How do they compare with the employment situations in Belfast, Blackburn?

Although the figures show something of a decline in manufacturing industry in the period, East Anglia had the second fastest growth of output in that period, after Wales. This has largely been the result of computer technology, which has demanded a smaller workforce, but a more highly skilled one. Thus the region has seen a burgeoning of the so-called 'high-tech' industries, with particularly rapid growth in the area around Cambridge. The table shows that service industries grew rapidly in that period, particularly when compared to the rest of Britain. Employment in

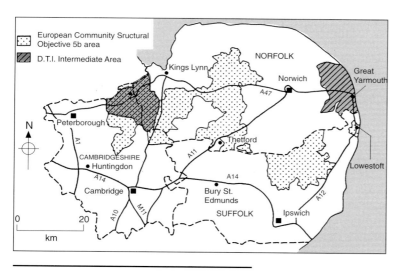

Figure 7.5.1 *Areas of regional aid in East Anglia*

	East Anglia		Change 1988–93		
	1988 '000	1993 '000	'000	%	Great Britain %
Agriculture, forestry, fishing	31	24	−7	−22.6	−19.3
Engineering and vehicles	75	65	−10	−13.3	−22.5
Other manufacturing	91	81	−10	−11.0	−13.3
Finance and business services	74	91	+17	+23.0	+7.0
Retailing, hotels, catering	161	173	+12	+7.5	+3.2
Education, health, other services	225	246	+21	+9.3	+3.5
All manufacturing industries	183	161	−22	−12.0	−17.6
All service industries	510	560	+50	+9.8	+3.0
Total economy	771	783	+12	+1.6	−4.1

Figure 7.5.2 *Changes in employment in East Anglia*

Figure 7.5.3   *New Industrial Estate, Peterborough*

financial and business services grew the fastest, and in a number of cases this reflected the decentralisation of employment in this sector from London. Four key factors would appear to have encouraged growth in both manufacturing (growth of output) and service (absolute and relative growth) industries:

- the shift of industry from urban locations to rural areas. Although this is something of a national trend it appears to be more clearly marked in East Anglia than in most other areas. Companies have relocated from London to the small market towns and other rural towns in East Anglia;
- the attractiveness of East Anglia as a residential location, particularly for middle and senior managers and professionals. Many have set up their own small and medium-sized businesses in the region;

- the benefits of relatively easy access to London and the South East, through new motorways, such as the M11 and the A1(M), and recently electrified rail links. This gives access not only to the major markets for goods and services in the United Kingdom, but also to the financial service centre of the country, and the major international communications focus of London;
- recent government designation of two assisted Intermediate Areas for aid focusing on Wisbech and Great Yarmouth. An Objective 5b area was created in parts of Cambridgeshire, Norfolk and Suffolk to receive EU financial assistance.

Within such an economically buoyant region, Peterborough, with its original New Town status has had the opportunity to flourish (Figure 7.5.3). Initial funding of Peterborough's expansion was through the New Towns programme, which was largely managed by the Peterborough Development Corporation. Just after the designation of the New Town, Peterborough still had nearly 50 per cent of its workforce employed in manufacturing industry, but this fell to barely 20 per cent by 1991. Service employment, as might be expected from national and regional trends, rose to nearly 75 per cent. Figure 7.5.4 shows the eleven most important sectors in employment in 1991. It shows an interesting mix of both manufacturing and service industries, with

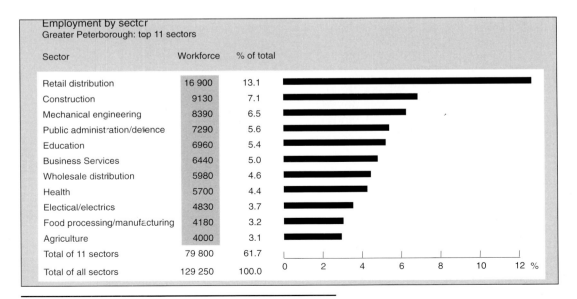

Employment by sector
Greater Peterborough: top 11 sectors

Sector	Workforce	% of total
Retail distribution	16 900	13.1
Construction	9130	7.1
Mechanical engineering	8390	6.5
Public administration/defence	7290	5.6
Education	6960	5.4
Business Services	6440	5.0
Wholesale distribution	5980	4.6
Health	5700	4.4
Electical/electrics	4830	3.7
Food processing/manufacturing	4180	3.2
Agriculture	4000	3.1
Total of 11 sectors	79 800	61.7
Total of all sectors	129 250	100.0

Figure 7.5.4   *Top 11 employment sectors in Greater Peterborough*

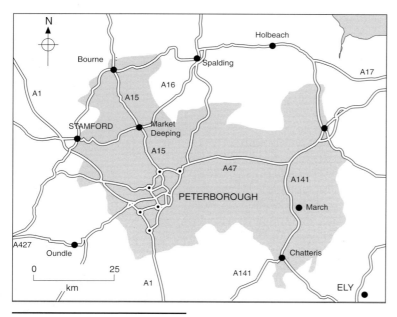

Figure 7.5.5 *Greater Peterborough*

mechanical engineering (largely the manufacture of diesel engines) holding up well amongst the growing service industries. In the service sector important employers include Barclays Bank, the Norwich and Peterborough Building Society, Pearl Assurance (headquarters relocated from London) and Royal Insurance.

Much of the management of the industrial environment in the Peterborough areas has now passed to the Peterborough Development Agency from the former Development Corporation. The formation of the Greater Peterborough Partnership, which covers a much wider area, and takes in the small towns of Spalding, Wisbech and March should ensure effective integration of both urban and rural prospects within the region (see Figure 7.5.5).

7.6 Managing a New Industrial Environment: Western Ireland

Traditionally, western Ireland has been regarded as one of the poorest parts of the Republic of Ireland and one of declining and outwardly migrating population trends. Work in the 1980s confirmed such a view, showing that rural deprivation was widespread in the west of Ireland. These are shown as 'rural problem areas' in Figure 7.6.1, which also shows the Less Favoured Areas of Ireland as defined in Directive 286 of the Common Agricultural Policy of the European Union. Current population trends in western Ireland show population declining in most areas, with some notable exceptions in the immediate hinterlands of Galway, Limerick and Cork.

Formal recognition of the need to bring some financial help to these areas was first recognised in 1952, with the creation of the Designated Areas, in which government grants were to be made available for small-scale industries. Although these grants were later extended to larger firms and the tourist industry, it does not appear that they were particularly effective in stimulating much-needed industrial growth in the west. In 1959, the establishment of the Shannon Free Airport Industrial Estate was an important step in the regional assistance programme.

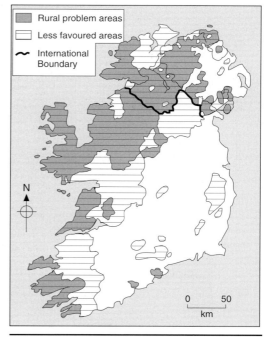

Figure 7.6.1 *Rural problems and less favoured areas in Ireland*

How Ireland's west hopes to be won

THEY have a soft spot in the west of Ireland for Albert Reynolds – the republic's temporary Taoiseach. Whatever else he might have done, the former Fianna Fail leader managed to convince the west – the heartland of the party's supporters – that he took their problems seriously.

Campaigners in the west, led by the region's bishops, are battling to maintain their communities in the face of emigration and migration. According to Bishop Thomas Finnegan of Killala, County Mayo, if the exodus continues, the population will shrink a further 20 per cent in under 20 years. "I see beautiful towns and villages, that survived the famine, now becoming ghost parishes," he said.

The aim of the crusade is to stem the tide by encouraging businesses – both local and overseas – to bring jobs to the west. Communications now make it possible for a host of service-type industries to locate there and the region is determined to win them over. Against such a background, Mr Reynolds's presumed support may not seem surprising. But many in the west believe that neglect by Dublin's establishment has a lot to do with their problems and, with the political situation in turmoil, they fear whoever emerges as the next Taoiseach may not prove as sympathetic.

Jim McGuire, a former journalist who is closely linked to the bishops' efforts, maintained: "If as much concentration went into saving the west of Ireland as saving their seats it would be fine. European structural funds allowed the country £7 billion but plans did not include the west, largely due to its underdeveloped nature."

There is also private concern over the impact of peace in the north. If the ceasefire holds, it is inevitable that some investment, which may otherwise have gone to the west, will go to the north although Mr Reynolds promised help for both.

But Lisa McAllister, chief executive of the Council for the West, formed six weeks ago to represent local interests, said: "The advantages of peace outweigh the disadvantages. People have to stop thinking like that. It's one, small island."

A report commissioned by the bishops earlier this year forced the Reynolds government into action. It called for a development board and estimated that IR£2 billion would be needed to tackle the crisis. The government set up a task force which came up with a plan to establish the Western Development Partnership Board and it is now working on a five-year plan.

It is a welcome move. According to Bishop Finnegan, sectoral divisions have led to such absurdities as plush hotels being built in out of the way locations with no account of the need to improve local infrastructure.

The trouble is that government only committed a budget of £100,000 for the board for 1995. There is also a suspicion that the board – with its majority of state-linked members – will block any initiatives that Dublin does not like.

MR McGUIRE said the task force was "quietly lured up an alley and mugged by the bureaucrats. The WDP is a watered-down version of what we wanted – autonomy."

To maintain the local momentum, the Council for the West was formed. The campaign is not solely directed at commercial interests as Ms McAllister explained: "It is not true that an improved economic situation follows on to a better social situation," she said.

Her main role is encouraging local people to make sure the issue does not die. Such campaigns have foundered on local apathy before and there is a real danger that official support is merely lip-service.

This time, though, the campaigners believe it will be different. According to Bishop Finnegan: "There is a feeling now that something can be done." He makes no apologies for the Church's involvement, although critics claim it should not be "meddling" in affairs of state. In an article last year, he wrote: "I do not accept that the Church deals only with an area of life that belongs to God and the state, with an area that belongs to man only . . . I am not prepared to preside in silence over a community that is breaking down before my eyes."

The council believes strongly that something can be done. Mr McGuire points out that the region is a prime fishing area, ripe for good quality tourism.

It is a view endorsed by the Irish Development Board – which has proved successful at attracting overseas firms there.

Tom Hyland, the IDA's Galway-based chief, said that he expects there to be "major investment" in hotels in the next few years and Jim Murren, also with the IDA, said: "The perception of remoteness here is unfounded."

Knock became something of a laughing stock when a local campaign saw an international airport built there. To describe it as a small village would be to overestimate its magnitude. But the airport is now at the centre of plans for redeveloping the west, according to Mr Murren.

There is also talk of building an international hospital, to attract rich Americans. Mr McGuire points out that Knock has an advantage over the recently failed Scottish hospital in that it can boast a shrine.

A clutch of international businesses, which the IDA has persuaded to move to the region, could not speak more highly of the province but financial incentives are the primary factor for them.

But will the firms run out when the incentives do? Oliver Cromwell threatened to drive the Irish to hell or Connaught – he considered the bleak province the best substitute on earth for the inferno. It might seem ironic that the bishops of the west are trying to preserve the population in the wild region, west of the Shannon . . . unless you happen to live there.

The Guardian 10 December 1994

Figure 7.6.2   *How Ireland's West hopes to be won*

Shannon Airport, once an important stepping stone on trans-Atlantic flights, was now increasingly being by-passed by longer-range jet aircraft, with consequent loss of revenue and local employment. A free trade zone was established and this led to the development of a range of industries on the airport estate. Although the estate created much-needed local employment in the Limerick area, it could not be regarded as a model for other developments in western Ireland simply because the circumstances at Shannon were so special and localised.

Following the publication of the Buchanan Report (see Chapter 6, page 113), the Industrial Development Authority was formed; this body was responsible for regional planning of industry in the whole of the Republic, except the mid-west, which was subcontracted to the company already running the Shannon Airport Estate. Although the targets set by the Board were often unrealistic, a general trend towards the ruralisation of industry became evident, and the west of Ireland seemed to benefit. However the relatively high growth levels in the early 1970s were not well sustained into the late 1980s. Inward investment from overseas has been an important ingredient in the development of industry in the West of Ireland. This has tended to be dominated

by transnational corporations based in the USA. Judged by the yardstick of the inherent problems of the remote and disadvantaged nature of the west, progress has certainly been made: industrial employment now stands at approximately 20 per cent of total employment in the west, and the west's share of total national industrial employment has risen steadily since the 1960s to reach approximately 30 per cent. To what extent these improvements will continue remains to be seen.

STUDENT ACTIVITY

1 Study the article 'How Ireland's west hopes to be won' (Figure 7.6.2).
   a What are the main hindrances to development in the West of Ireland?
   b What types of development does the campaign, led by the Council for the West, hope to develop?
   c To what extent do you think that the west of Ireland is more suited to tourism, and other service industries, rather than manufacturing industry?

ESSAYS

1 To what extent has the globalisation of industry aided the regeneration of old industrial areas?

2 Assess the merits of service industries as replacements for manufacturing industries.

FIELDWORK OPPORTUNITIES & PROJECT SUGGESTIONS

1 Select an industrial estate, on which a variety of industries are located (these will include both manufacturing industries and service industries). Draw up a classification of both manufacturing and service industries, which you think will cover most types on the estate (you may have to modify it after your fieldwork). Obtain a map of the estate (usually available from the local authority, or the estate manager), and then plot the industries according to your classification, and plot all the vacant buildings also. Select a range of criteria which will enable you to assess the environmental quality of the estate, e.g. quality of buildings, amount and

quality of lawns and trees, parking facilities. Management initiatives could then be mapped, e.g. security fencing, traffic calming, signposting. You could then compare this estate with a different one, e.g. a local authority estate compared with a private estate.

2 Choose two different areas of a town where office functions are important. It might be possible to compare office functions in the city centre, with those in a new office park, or business park on the city fringe. Classify the types of office function in the two different areas, and attempt to find out if there is any clustering of one particular activity in either of the two areas.

Managing Rural Environments

8

KEY IDEAS

- Apart from the most remote areas in Britain and Ireland, counterurbanisation has seen an increase in population in rural areas
- An increasing gap is appearing between the affluent and the underprivileged in rural areas
- Decline in rural services particularly affects deprived groups
- A key management role is the provision of essential services in rural areas
- Strategic planning and management of rural areas seek to maintain a balance between different districts within an integrated whole

8.1 Introduction

Lord Shuttleworth's comments summarise some of the important issues that now face those who are responsible for the management of rural environments in Britain and Ireland. Although the nature of rural areas varies widely across the United Kingdom and Ireland, common themes link areas as diverse as the far west of Ireland and the leafy commuter lands of the Surrey Weald, the windswept uplands of the Northern Pennines and the arable expanses of Lincolnshire. 'Jobs, homes and services' are mentioned in the quotation from Lord Shuttleworth's speech. We might add leisure and recreation in the countryside, and conservation of a valued landscape. We then have a vision of living communities set in a rural environment with much worth protecting, yet which is also readily available for people's enjoyment. The main issues in rural management are shown in Figure 8.1.1.

STUDENT **8.1** ACTIVITY

Study the article 'Only three rural idylls elude the madding crowd' (Figure 8.1.2).
1 Why should we need to preserve the 'godly quietness of the countryside'?
2 What is the nature of the threats to the 'tranquil' areas?
3 How realistic are the measures proposed to deal with the declining areas of 'quiet land'?

In November 1994, the *Guardian* asked six writers to comment on the changes that had occurred in places that they knew as children. Three of the accounts by the writers are reproduced, in abbreviated form, below (Figure 8.1.3). One is on the urban fringe of the countryside – the Sankey Brook in Lancashire (Sir John Johnson), one is near the coast in Kent (Marion Shoard), the third is a village on the border of Oxfordshire and Gloucestershire (Tracey Worcester), probably the most rural of the three.

'Without policies which recognise and enable rural areas to be places which provide jobs, homes and services, for a mix of people in differing circumstances, then we will see a decline in rural communities and a loss of much of what we value and enjoy in the countryside.'

Lord Shuttleworth: House of Lords' debate on rural policies, 10 March 1995.

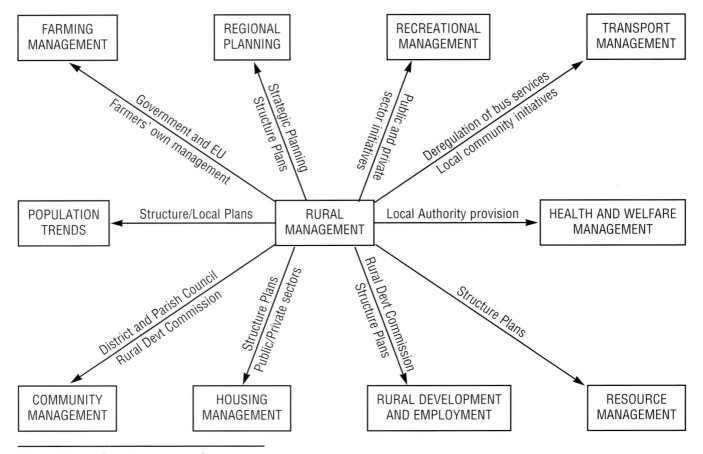

Figure 8.1.1 *The main issues in rural management*

Only three rural idylls elude madding crowd

RURAL idylls are heading for extinction. Once refuges from the 'madding crowd's ignoble strife', today they reverberate to the sound of yuppies playing war games, clay pigeon shoots, banger racing and the thunder of passing juggernauts.

This week the Council for the Protection of Rural England publishes a report, *The Tranquil Area*, which says new roads, housing estates, industrial parks and 'leisure' activities such as car racing are fast destroying the countryside's 'godly quietness'.

It reveals that only three 'pockets of peace' – north Devon, the Shropshire and Herefordshire Marches, and the north Pennines – have survived the past two decades of development. The rest is sound and fury.

'Those areas that remain relatively undisturbed have shrunk to a quarter of their previous size,' says Tony Burton, a CPRE senior planner. 'An area of calm the size of Wales has been wiped out.'

The CPRE criteria stipulated that only land unaffected by power stations, motorways, other major roads, airfields or open-cast mines can be called 'tranquil'. This definition revealed that South-East England has virtually no quiet areas left; the South-West, once the most tranquil part of England, is breaking up faster than any other area; and the Midlands and East Anglia are badly affected.

Berkshire provides a classic picture of this havoc. Bounded by the M25 to the east and divided by the M4, it has lost 45 per cent of its tranquil areas over the past 20 years. It has new towns, estates and service stations. Even its gravel quarries, created to provide road-building materials, have been flooded and are used for water

scooter riding and other noisy sports. 'Take my village, Aldermaston,' said Commander Michael Porter, the CPRE's Berkshire chairman. 'Over the past few years, at the edge of the village, people have started playing war games in the local woods. We have clay pigeon shoots and enthusiasts have begun racing old cars.'

The report urges the Government to adopt key measures to protect the last of the nation's quiet refuges. These include:

- Implementing urban regeneration schemes to make city life more palatable and halt urban depopulation.
- Investing in rail rather than road transport.
- Halting all schemes to build new towns or estates.

The Observer

Figure 8.1.2 *The rural idyll*

STUDENT **8.2** ACTIVITY

1 Which of the three has undergone the most change?
2 To what extent have the changes been for the better, or for the worse?
3 How do the changes reflect some of the incursions into the countryside mentioned in the article 'Only three rural idylls elude the madding crowd'?
4 What rural management issues are involved in each of these three accounts?

In Britain and Ireland, broad guidelines for rural management are embodied in both European and individual Government legislation and directives. Initiatives, such as the designation of the Less Favoured Areas under Directive 286 of the European Community (see Chapter 7, page 136) have brought aid to farmers living and working under difficult physical conditions in the upland areas of both Britain and Ireland. At Government level, initiatives such as the White Paper on Rural England, published in October 1995, act as a blueprint for future development. It was a

Change in rural areas

Marion Shoard

Author of The Theft Of Our Countryside and This Land Is Our Land

NO SMELL is more evocative for me than that of fennel. Break a stalk of this tall, yellow-green plant and its aniseedy whiff transports me at once to the cliffs above Pegwell Bay in east Kent, snug under the chin of the Isle of Thanet. A child once more, I am squatting in a jungle of wild parsley, fennel and alexanders, staring down at the shining circle of mud and sand which ties the chalk hills of Thanet to the saltings at the mouth of the River Stour. As a family living in Ramsgate, one of our favourite outings was to Pegwell, clattering up the stairs of the double-decker bus, pouring off near Pegwell village, then taking the cliff-top path. We would search for delicate pink Venus shells and old ships' timbers on the shore below before setting out on the wooden causeway across the calf-deep black mud to firm, wide, muddy flats where the sea lapped invisibly a mile away at low tide.

In the 1960s, qualities which had attracted the early voyagers drew a new breed of seafarers. The imperceptible slope which had proved so welcoming to the shallow-draught craft of the Romans and Jutes was deemed ideally suited to what appeared to be the longships of the future. In 1969 a central section of the cliff was hacked out and a hovercraft terminal was built slap in the middle of the Bay's

Tracey Worcester

Environmental activist

CORNWELL, the village on the border of Oxfordshire and Gloucestershire where I was brought up, provided an idyllic childhood. I played in the river which flowed around the cottages, across the village green, past the large stone Manor House, and on over a crayfish-filled waterfall. The adults knew all the children and kept an eye out for them.

The manor and village were built by a sheep farmer for his staff and labourers in the 16th century. Not much has changed since, except that part of the farm has been leased to a tenant who also lives in the village. The air is clean and not one child has asthma, possibly because the single road into the village goes nowhere else and has managed to escape the thousands of miles of "improvements" made since my birth in 1958. The stream where I played rises in a field near the cottages and its water is pumped to them untreated, direct from where it bubbles up.

No pesticides have ever been used on the parkland through which the stream flows, so there are still plenty of aquatic insects to feed the fish. They in turn feed the electric blue kingfisher, which has disappeared throughout much of this country. With no nitrogenous fertilisers used, the fish have not been poisoned by algal blooms.

Sir John Johnson

Chairman of the Countryside Commission

MORE than two centuries ago there was a wooded valley in Lancashire down to the Mersey. Our ancestors did their best to destroy it in the name of industrial progress. But, when I knew it 50 years ago, it was my countryside. It was distinctive and it had two special attractions: the Stinking Brook and the Mucky Mountains.

No one called these features anything else, although on the map one was the Sankey Brook. It was a grey broth of chemical foam and nauseous mud. Beside it was the Sankey Canal, one of the earliest arteries of industrial Britain.

The Mucky Mountains were a wilderness, a miniature range of hills with a definite summit, stunted bushes and grassy hollows. Here was a recreational facility tailor-made for wide games and testing cycling, long before mountain bikes were thought of. It was an alien landscape, a calcareous spoil heap from an old alkali works which processed bargefuls of salt and stone brought to the canal wharf below.

The whole area has been transformed in the past 20 years. Local government reform in 1974 brought two new authorities: Merseyside County Council and the Metropolitan Borough of St Helens. In addition the

Change in rural areas *cont.*

shore. Its stark, brutalist, shape immediately dominated its surroundings. Soon, its howling craft were lifting their skirts over the mud and speeding out over the Bay, scattering the oystercatchers and curlews.

The commotion was fleeting. After 13 years of operation the hoverport closed. But the Bay has not recovered from its brief life as a gateway to the future. The 20-acre terminal site lies derelict. Smashed windows look down on cracking tarmac. The vast car-park is empty except for a handful of container lorries in one corner and a mass of black tyres making a go-kart track in another. Rubbish is dumped on the edge of the site. The brooding monument to the fickleness of commerce sulking in Pegwell's heart makes the Bay now an unsettling place to linger.

Cornwell has escaped the fate of so many villages which are virtually ghost communities during the week; their cottages sold or rented to weekend commuters. In Britain as a whole, 12,000 small farmers and farm labourers are leaving the land each year as expensive hi-tech farm machinery and imported food render their holdings uneconomic. The village has survived untouched because the owner of its 1,800 acres uses the profits from a successful brick business to sustain the high labour force.

1980s saw the development of St Helens' Groundwork Trust. Together these bodies saw the Sankey corridor as ripe for improvement. A grey wedge running through dense conurbations could be a green way.

Nature did its bit. The end of old industries meant no more dumping of spoil and less chemical pollution. The residue of an age of adverse landscape change was left to itself. Wetlands were recreated and attracted a wealth of bird life. Clumps of bushes by the dried canal bed harboured migrant owls from Scandinavia in the winter. The Mucky Mountains became a site of biological interest. Seeds blown in from the coast colonised the unusual limestone habitat and orchids grew.

Here was a lesson in what people could do with their environment. A network of footpaths was threaded through the valley and its slopes. Trees were planted. Local communities took pride in the new habitats and made good use of their landscape. The Sankey Canal Restoration Society, a voluntary body, grew from small beginnings into an enthusiastic force and worked to re-open the canal to navigation.

Much of the area is now covered by the Mersey Community Forest. The aim is to develop a patchwork of woodlands, farming, wildlife habitats and areas for sport and recreation. In 50 years the old industrial sites will have been redeemed.

The Guardian

Figure 8.1.3   *Britain's changing places*

document that was generally well received by politicians, farmers and conservationists alike.

Management at Local Authority Level is largely through the planning framework established in County Structure Plans, and District Council Plans within the United Kingdom. In Northern Ireland the Planning Strategy for Rural Northern Ireland is a document that covers rural policy in the whole of the Province. In the Republic of Ireland physical development of the rural areas has been largely controlled through the Planning and Development Acts of 1963, 1976 and 1982. However, failure to implement the policy of concentrating non-farm housing near existing settlements has resulted in the so-called 'bungalow blitz' (Figure 8.1.5) so characteristic of some of the most scenically attractive areas in the west of Ireland.

In areas where conservation is important, such as National Parks and Areas of Outstanding Natural Beauty in the United Kingdom, and the National Parks in the Republic of Ireland, management policies specific to these areas are superimposed on those already in existence through local authorities. Additionally bodies such as the Rural Development Commission (Figure 8.1.4) provide a further, and important tier to countryside management.

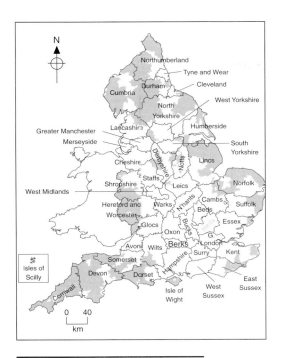

Figure 8.1.5   *Bungalow blitz', Achill Island, Ireland*

Figure 8.1.4   *Rural development areas*

8.2 Case Studies in Rural Management

The case studies are prefaced with a review of changing population patterns in rural Britain and Ireland. Although depopulation is still a trend in the more remote parts of western Ireland, and parts of Scotland and mid Wales, counterurbanisation has been responsible for an increase in population not only in the rural areas adjacent to the urban areas of the countries, e.g. in the Home Counties, and in the areas of central Ireland to the west of Dublin, but also in some more remote areas.

Rural deprivation has emerged as a basic problem in many rural areas, and it is considered next, both in general terms, and in the context of 'Lifestyles in Rural England'. Two elements in rural deprivation – housing and mobility are then considered, each an important management issue.

Figure 8.2.1   *Western Purbeck, Dorset: still relatively unaffected by counterurbanisation*

8.3 Recent Population Change in rural Britain and Ireland

Although loss of population has long been a characteristic of some rural areas in Britain and Ireland, the recent population census of 1991 indicates that the trend towards an increase in population in rural areas is a significant one, although perhaps less so in the Republic of Ireland. Much of this growth reflects the trend known as counterurbanisation. This phenomenon is not only characteristic of rural areas close to towns and cities, but is also becoming noticeable in more remote areas. A number of factors are responsible for this growth of population in rural areas:

- increasing levels of car ownership, making living in the countryside, but working in the town, more possible;
- rapid growth in communications technology, which makes working from home a viable proposition for increasing numbers of people;
- the urban–rural shift of industry (see Chapter 7, page 135);
- improvements in the quality of life to be gained from living in the countryside;
- the increasing numbers of relatively affluent people of retirement age who seek the attractions of living in some of the more remote rural areas.

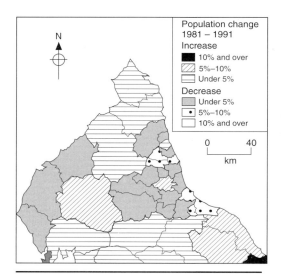

Figure 8.3.1 *Population change in part of northern England, 1981–91*

The map of southern England (Figure 8.3.2) shows an almost universal increase in population between 1981 and 1991, except in the large towns and cities and the areas adjacent to them in the Home Counties. Even relatively remote areas such as north Cornwall are shown to

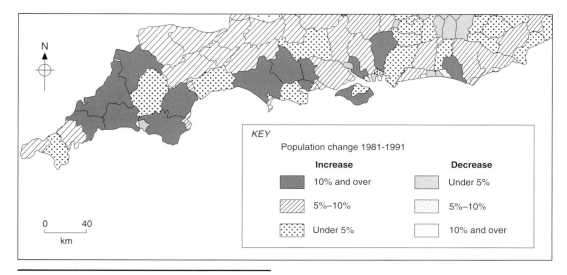

Figure 8.3.2 *Population change in southern England*

Figure 8.3.3 *Abandoned farm buildings near Ribblehead, north west Yorkshire*

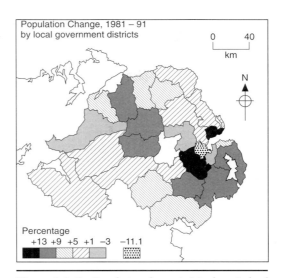

Figure 8.3.5 *Population change in Northern Ireland*

experience growth levels of 10 per cent and over, thus comparing favourably with parts of the much more accessible East Hampshire and Sussex. The map of part of northern England (Figure 8.3.1), on the other hand, shows something of a different pattern. Although there has been a more modest growth of population in rural areas, the more remote areas in Cumbria and the Pennine Dales, still show a loss of population (Figure 8.3.3).

The map of Scotland (Figure 8.3.4) illustrates population change between 1981 and 1991. Some rural areas show a significant increase in population, such as the Dee valley running inland from Aberdeen, and the Island of Skye, and the neighbouring mainland area. Other more remote areas, such as Sutherland and Caithness and considerable swathes of the Southern Uplands have shown an overall loss in population. The map of Northern Ireland (Figure 8.3.5) shows only one area of declining population, in the remoter west of the Province, in County Tyrone. The map (Figure 8.3.6) shows, in the Republic of Ireland, a somewhat mixed

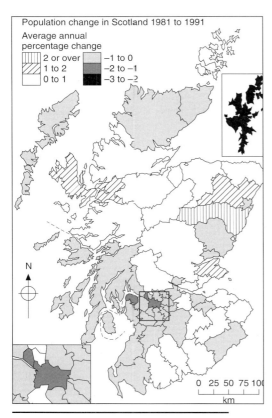

Figure 8.3.4 *Population change in Scotland, 1981–91*

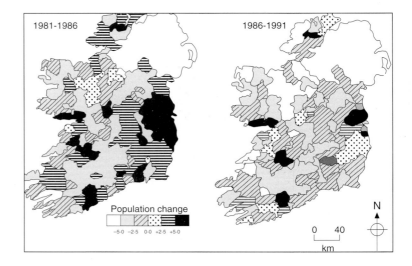

Figure 8.3.6 *Population change in the Irish Republic*

pattern of growth and decline. As expected, the phenomenon of counterurbanisation is clear enough in the areas of population increase around Dublin, and to a lesser extent around the second and third towns of the Republic, Cork and Limerick. Vast tracts of the west and centre, however, recorded a loss of population in the period between 1986 and 1991. Interestingly, this marks a sharp change from the early 1980s, when many rural areas were still recording a growth in population. The decline in the late 1980s, likely to continue into the 1990s, seems to result from a renewed surge of migration from the countryside, and the first signs of a reduction in fertility consequent on social change in rural areas.

STUDENT ACTIVITY

1 Using the appropriate maps in section 8.3, discuss the reasons for change in rural population in the 1980s (or late 1980s in the Republic of Ireland), in the following areas
   a south west England
   b a cross section of northern England, from Cumbria to the lower Tyne Valley
   c the Highlands of Scotland
   d the Irish lowlands to the west of Dublin
   e the west coast of Ireland.
2 Explain the reasons for the differences that you note between the five areas.

Managing the Problems of Living and Working in Rural Environments

STUDENT ACTIVITY

1 Read the extracts from the article 'A terrible kind of beauty' (Figure 8.4.1).
   a List the main problems of rural living identified in the article.
   b What are the main management issues that arise from these problems?
   c What should be the main management objectives in tackling the problems?

The existence of rural deprivation has been an important theme in the study of rural living in the second half of the twentieth century. Howard Newby wrote in *England's Green and Pleasant Land* – 'It is easy to overlook these problems amongst the general prosperity of contemporary rural England. The appearance of many villages suggests two car families enjoying a lifestyle of comparative affluence in their beautifully restored houses. The other face of rural England is more difficult to seek out since it is less openly admitted . . .

poverty brings with it a sense of exclusion . . . The rural poor find that they and their needs are increasingly regarded as residual or even unacknowledged . . . Consequently their inclination is to make do, while the general public is given little reason to alter its image of a cosy and contented countryside.'
Three main aspects of rural deprivation are now recognised:
● household deprivation: problems related to income and housing which dictate the ability of rural dwellers to use the opportunities that are available to them;
● opportunity deprivation: problems related to the loss of jobs and services from rural areas;
● mobility deprivation: problems stemming from the difficulty that some rural people have in gaining access to employment, services and facilities that are no longer available in the area where they live.
The combined effects of these three facets of deprivation serve to create very

A terrible kind of beauty

IN PICTURES and photographs, rural life can still look idyllic: peace, calm, natural beauty, breathtaking landscapes. But behind the pictures life is different – not just for the poor, but for the middle income too.

The latest national count shows 2 in 5 villages in Britain have no shop, 1 in 3 has no post office, 3 in 5 no school, 3 in 4 no GP. More than half the workers in farming, forestry or fishing earn less than £200 a week. A car is often indispensable for mobility, but a quarter of rural households don't have one. Suicide rates are particularly high in the countryside.

Poverty affects between 20 and 30 per cent of people living in the British countryside. It is usually hidden from view, but it is real, with low pay, isolation and an increasing lack of public services among its starkest characteristics.

Poverty is not a word that government ministers willingly use, and it can take policy-makers and even backbench MPs a long time to come to terms with its existence. But as Les Roberts, director of Acre, Action for Communities in Rural England, says: "You can't put up a notice in the village hall saying: 'Will all poor people come to a meeting?' People who are poor may be ashamed and want, above all, to keep their dignity."

The attributes of poverty in the countryside are different from those of the urban area, as are the steps that have to be taken if it is to be alleviated. Things taken for granted in the town or city are conspicuously absent in the countryside.

The picture is not one of unrelieved gloom. In many areas, local workers have made significant inroads. In Derbyshire, a take-up campaign for available benefits for the poorest has made noticeable progress; in Hereford and Worcester, debt support groups are developing; in East Sussex, skill exchange arrangements are taking off; and in rural Durham, training programmes, with built-in child care facilities, are growing. For Acre, the prime concern is to keep poverty on the political agenda. "At least," says Roberts, "it is no longer ignored."

Acre, set up 10 years ago, works on a necessarily limited budget, but draws its strength, and clout, from its unique network of 38 rural community councils. One of its concerns is not to get swamped by the enormity of its workload. In his latest annual report, Roberts warned that resources were easily drained in the direction of national issues and concerns, and that Acre could not afford to be drawn into long debates over the flavour of the month.

Despite personal and professional commitments, Roberts, a former social worker, says he does not wish to be seen as part of what he calls "the poverty industry".

On page seven of the annual report there is a cameo of Bill Rider, a craggy-faced, retired Devon farmer, who has the tab end of a home-rolled cigarette on his lips and two days' stubble on his chin. He has confronted changing social patterns in the countryside head-on and contemplates them today without a hint of nostalgia – but also without any nonsense.

"Before the war," Rider says, "the village of Blackawton had a blacksmith, a boot repairer, two grocers, two butchers, a post office and a baker. Now there's only one shop and a post office."

In all parts of the country, social workers and others working with the rural poor talk of the extra costs that confront them. Travel and subsistence costs for staff are higher because client families live at varying distances from the work base; office rents and security risks may be higher than in the town; and home visits are less frequent than the workers would wish because of the distances involved.

Identifying the "target group" can also present problems which do not arise so often in the urban environment. In Nottinghamshire, one of England's poorest counties, social workers have problems not only in identifying pockets of poverty, but also of bringing people together to discuss common problems.

"Not only are people unwilling to acknowledge there is poverty and deprivation in their village or area," says Godfrey Claff, a liaison officer in the Peak District of neighbouring Derbyshire, "but they disguise it, too. A lawyer married to a teacher may live in a remote cottage, next door to a jobless couple on benefit who are just as smartly turned out."

At Aberdeen University's department of land economy, Professor Mark Shucksmith, who has examined poverty in rural Scotland, talks of people who are "rich in spirit, but poor in means" – people who insist they don't *feel* poor and who are often reluctant to take up state benefits to which they are entitled. He tends towards the EU definition of poverty, which centres on exclusion and powerlessness.

Finally, there is the problem of the non-poor being unwilling to accept that there is in fact a poverty problem. At one level, this shows in town dwellers who move out to live in the countryside and won't, according to the specialists, allow their rural vision to be shattered. At another level, it shows in a village doctor who railed at one of his patients for publicly declaring that poverty was an issue in their community.

Meanwhile, it remains a British trait to make much of its countryside. In an average year, 3 in 4 people visit the countryside at least once a year – making around 600 million "visits" in all. "I sometimes wonder," says Les Roberts, "whether they really know what they are looking at."

The Guardian 28 February 1996

Figure 8.4.1   A terrible kind of beauty

considerable problems in achieving a decent standard of living for certain groups within the rural community. As Newby points out the incidence of this deprivation very often remains hidden within the community, and does not command the attention that its urban equivalent is accorded.

In an important study in the 1980s McLaughlin found that in his selected areas (parts of Essex, Northumberland, Shropshire, Suffolk and Yorkshire)

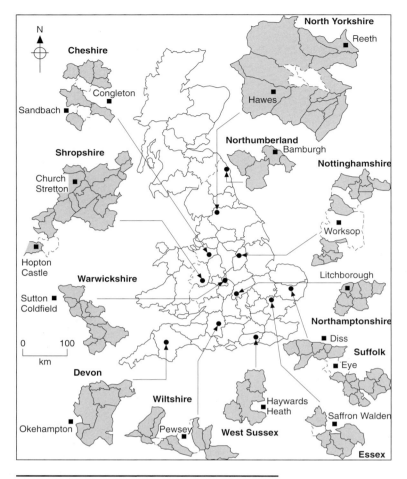

Figure 8.4.2 *Survey areas for lifestyles in rural Britain*

deprivation affected some 25 per cent of those households that he questioned. The predicament of the deprived was seen in low incomes, low levels of personal mobility and consequent problems of access to good quality housing and to a range of public and private services.

A further valuable study – 'Lifestyles in Rural England' was carried out in 1990. Whereas McLaughlin sampled five areas, the new study embraced all of McLaughlin's areas, but added a further seven (see Figure 8.4.2). Demographic, social and employment statistics for the 12 areas are given in Figure 8.4.3.

STUDENT **8.5** ACTIVITY

1 Attempt to rank the social status of the different areas by means of a composite index. Rank only those items which you think would be good indices of social status. Then derive a composite index, by summing the different rankings for the items that you have selected.
2 From your ranked list choose three areas that are obviously different in social character, and that are also, if possible, from different parts of England.
3 For each of your selected areas, use all of the statistics to draw up a list of rural management issues which are likely to be significant.

	Age			Tenure				
	under 16 %	16–64 %	65 and over %	owner occupied %	private rented %	social rented %	in non-perm. accom. %	recent movers %
Cheshire	17.2	68.3	14.5	67.8	20.0	12.2	1.5	6.6
Devon	20.7	60.1	19.2	76.2	13.1	10.7	0.9	8.0
Essex	22.7	66.4	10.9	68.2	19.6	12.3	0.2	10.6
Northamptonshire	20.0	65.9	14.1	78.1	9.5	12.4	0.4	8.9
Northumberland	16.0	59.1	24.8	59.6	17.6	22.8	0.8	9.6
North Yorkshire	17.4	62.3	20.3	73.1	17.6	9.3	0.3	6.2
Nottinghamshire	19.8	65.4	14.9	69.1	9.3	21.7	0.2	7.3
Shropshire	17.8	64.9	17.3	74.7	21.7	3.5	2.4	8.0
Suffolk	18.5	64.9	16.6	68.8	19.0	12.3	1.1	10.9
Warwickshire	17.5	68.6	13.9	79.1	13.8	7.0	0.7	5.9
West Sussex	17.2	63.7	19.1	67.6	18.7	13.8	8.0	10.6
Wiltshire	20.8	61.5	17.7	65.7	18.5	15.9	0.2	10.6
ENGLAND*	20.1	61.3	18.7	67.6	12.6	19.8		

	Car Ownership		Employment		
	no car %	two or more %	economically active adults %	unemployment rate %	self employment %
Cheshire	10.4	51.7	63.4	5.6	30.7
Devon	16.9	31.9	56.9	5.8	34.9
Essex	13.1	45.4	67.4	4.1	16.8
Northamptonshire	12.3	51.5	65.4	4.6	21.9
Northumberland	30.3	17.8	54.0	5.6	26.5
North Yorkshire	19.7	30.0	59.5	1.7	40.8
Nottinghamshire	28.9	32.3	59.6	9.4	13.8
Shropshire	8.0	45.6	59.6	6.0	37.7
Suffolk	14.1	41.9	60.1	4.0	24.7
Warwickshire	10.2	53.1	68.1	4.7	21.1
West Sussex	16.4	41.1	60.4	5.5	18.4
Wiltshire	12.4	45.1	61.0	3.8	22.4
ENGLAND	32.4	23.9			

Figure 8.4.3   Demographic social and employment statistics for 12 selected area in rural England

8.5 Rural Housing Problems and Their Management

In the above study three main issues emerge in a review of rural housing:

- there are pressures on restricted housing markets from affluent in-migrants who are capable of out-bidding local people for existing properties;
- there is a need for specific provision for affordable housing to rent, both to meet local needs, and to compensate for the cumulative shortfall, caused by lower rates of new building by housing associations than were previously achieved by local authorities.
- there has been an overall reduction in the social housing stock because of the right-to-buy policy.

Counterurbanisation seems to have had a substantial effect on the housing market in rural areas, both in those relatively near to towns and cities, and in the more remote areas. Restrictive planning policies have often led to a limited supply of housing, for which there is uneven competition between affluent newcomers, and less well-off local people. It is these original dwellers who have thus been forced into seeking affordable housing to rent, rather than compete unequally on the open housing market.

The present lack of affordable housing in rural areas has put particular pressure on young people. With a reduction in the stock of social housing, young couples are finding it impossible to compete in the owner-occupied housing sector, and are therefore often forced to leave rural areas altogether, and live in nearby towns. This inevitably leads to a loosening of family ties, part of the essential bonding of rural society.

Figure 8.5.1 shows a model of rural housing opportunity. Housing tenure occupies a pivotal position in the diagram. It is seen as being very much a function of household income, and, in its turn has an obvious influence on the quality and type of housing, and on the quality of life of its inhabitants. Three main types of tenure have been recognised: owner-occupation, local authority housing for rent, and privately rented housing. The advent of the housing association as the main provider of affordable housing has added an important fourth dimension (see

'They think they own the place, these people from the large towns. There are very few original villagers left now. They don't want to know original villagers. The town people come in and take over everything. They have put the price of homes up so that natives have to move out.'

Correspondent quoted in Lifestyles in Rural England, 1994.

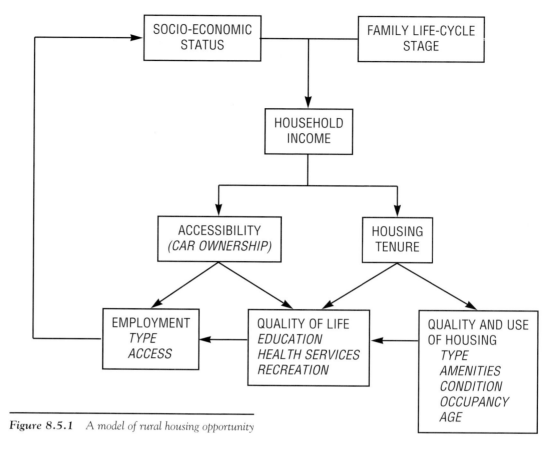

Figure 8.5.1 *A model of rural housing opportunity*

photographs, Figure 8.5.2).
The proportion of housing in each tenure category in the twelve study areas in 'Changing Lifestyles in Rural England' is shown in Figure 8.5.3. Roughly three-quarters of the housing stock is owner-occupied, although relative cost of this housing varies considerably across the twelve areas (Figure 8.5.4). It will be noted that the opportunities for entry into this sector of the housing market vary enormously across the country, being particularly difficult in counties such as

Figure 8.5.2b *Private Housing, Corfe Castle, Dorset*

Figure 8.5.2a *Local Authority Housing, Corfe Castle, Dorset*

Figure 8.5.2c *Affordable Rural Housing, Raglan Housing Association, Abbotsbury, Dorset*

	Owned – outright	Owned – mortgage	LA or HA rented	Private rented	Other
Cheshire	31.8	46.7	10.3	8.1	3.1
Devon	40.2	36.9	11.6	10.1	1.2
Essex	26.5	48.9	14.6	8.9	1.1
Northamptonshire	30.3	53.1	12.6	4.0	0.0
Northumberland	37.5	24.6	17.2	18.9	1.8
North Yorkshire	50.4	25.4	6.8	16.6	0.8
Nottinghamshire	28.0	38.4	25.2	6.8	1.6
Shropshire	57.1	25.3	3.7	13.2	0.7
Suffolk	41.6	29.4	13.0	14.9	1.1
Warwickshire	33.3	46.8	6.0	12.7	1.2
West Sussex	31.4	42.9	7.7	17.2	0.8
Wiltshire	33.9	31.9	20.6	12.8	0.8
mean for all areas	36.9	37.4	12.4	12.1	1.2

Figure 8.5.3   Household tenure

Warwickshire and Northamptonshire. Local authority housing plays a relatively unimportant part in the tenure mix in the twelve areas studied. In general, local authority housing has been less important in rural than in urban areas, and has become even less so with the current government constraints, and the loss of houses from this sector through the 'right-to-buy' policy. Housing associations, funded by grants from the Housing Corporation, have, in many areas, now replaced local authorities as the main source of affordable housing for rent. Although housing associations have yet to make a significant impact on the rural housing problem, there does seem to be some optimism for the approach which targets certain groups in the areas where housing shortages are acute. Privately rented accommodation is often related to a particular country estate that owns houses within a village: elsewhere the National Trust can be an important landlord locally. Tied cottages are rented by farm workers from the estate for which they work. They are important elements in the rural housing mix in the south west and the south east of England, and also in Scotland, where up to 80 per cent of agricultural workers are housed in tied cottages.

From a management standpoint, the most pressing issue in rural areas is the provision of enough suitable affordable housing. In 1993–4, the Housing Corporation's target for its rural programme was 1850 homes, about 6 per cent of its total programme.

Under £45,000		Over £105,000	
Northumberland	65.4	West Sussex	40.1
Shropshire	58.8	Warwickshire	38.3
Nottinghamshire	55.4	Northamptonshire	33.9
Suffolk	52.9	Wiltshire	32.4
Wiltshire	51.2	Essex	29.9
West Sussex	48.9	Shropshire	29.9
Cheshire	48.1	Cheshire	29.6
Essex	47.6	North Yorkshire	27.8
Devon	46.8	Devon	24.4
North Yorkshire	46.8	Suffolk	20.6
Northamptonshire	42.4	Nottinghamshire	0.4
Warwickshire	36.4	Northumberland	4.7

Figure 8.5.4   Properties thought to be worth under £45 000 and over £105 000

This has to be set against the Rural Development Commission's estimate that the rural housing stock needs to increase at a rate of 16 000 homes a year for the next five years. Rural housing schemes tend to be more expensive to build because of their scale, remoteness, and the need for high standards of design in the countryside. Generally there is evidence that developments are being skewed towards rural towns and large villages, rather than those smaller villages (<100 inhabitants) where the need is greatest.

STUDENT 8.6 ACTIVITY

1 Explain why counterurbanisation has unbalanced the rural housing market.
2 The greatest need in rural areas is for more affordable housing, particularly for rent. Suggest a five point action programme that needs to be adopted by the government to fulfil this need. What commercial and political pressures will make this difficult to realise?

8.6 Rural Transport and the Accessibility Issue

'In the countryside, private transport is now the key to maintaining the rural quality of life'

White Paper: Rural England (1995).

'However, many people in the countryside do not own a car. For these people there is a need for affordable, local transport services.'

Lord Shuttleworth, Chairman of the Rural Development Commission.

These two comments polarise views on rural transport. On the one hand, ownership of a car means that all of the advantages of living in the countryside can be enjoyed, without suffering any of the handicaps that greater remoteness, and reduced accessibility bring: quality of life is much enhanced. Lack of access to private transport means mobility deprivation, and all of the hardships that it brings: quality of life inevitably suffers.
In countryside areas, a higher level of car ownership would be expected, because of the higher levels of mobility that it confers.

The 1991 census suggests that 67.6 per cent of all households in England had a car, and 23.9 per cent had two or more. It would be expected that levels of ownership in the countryside would almost certainly be higher, in the region of 70–80 per cent. The 1991 census for the twelve areas used in the Lifestyles in Britain survey shows that in those twelve areas 83.9 per cent of households had access to at least one vehicle. Regional variations are shown in Figure 8.6.1, with only Northumberland and Nottingham falling below the 75 per cent level.

	No car	One car	Two or more cars
Cheshire	10.4	37.9	51.7
Devon	16.9	51.2	31.9
Essex	13.1	41.5	45.4
Northamptonshire	12.3	36.2	51.5
Northumberland	30.3	51.9	17.8
North Yorkshire	19.7	50.3	30.0
Nottinghamshire	28.9	38.8	32.3
Shropshire	8.0	46.4	45.6
Suffolk	14.1	44.0	41.9
Warwickshire	10.2	36.7	53.1
West Sussex	16.4	42.5	41.1
Wiltshire	12.4	42.5	45.1
mean across 12 areas	16.1	43.3	40.6
England	32.4	43.7	23.9

Figure 8.6.1 *Number of cars in household, 1991*

With such high levels of car ownership in rural areas, the number of potential passengers using public transport has been much reduced. This has inevitably meant a curtailed service, or higher fares, or both. The country bus network was densest in the 1950s, and since then there has been a steady decline in the numbers of people using the services. Between 1965 and 1975, there was a 30 per cent decline in passenger-miles, but only a 10 per cent decline in vehicle-miles. Hence buses were being used far less and became more expensive to run. Since 1968, local authorities have been able to subsidise some rural services, and with **deregulation** of bus services in 1986, the local authority has no longer had control over commercial operators, provided that they run the services that they have registered. The subsidised network acts as a social back-up to the commercial services, and has to be put out to tender by the local authority, which specifies routes, levels of service and fares. Figure 8.6.2 shows the bus network in north Powys, before and after deregulation. The much reduced commercial network after deregulation had to be supplemented by subsidised services operated under the broad control of the local authority. These latter services virtually restored the network to its earlier density.

Where neither the commercial operator or the subsidised service provides adequate route coverage, then self-help in the local community is often the only answer to providing a socially necessary service. Community bus services, run and operated by volunteers are one way forward, and the multi-purpose service, such as the Post Bus is another. Many rural areas now have a mix of commercial, local authority, and voluntary services but this is far from being universally available.

In the early years of the twentieth century, many villages, and even hamlets had their own station or halt on a railway network that one writer described as 'almost absurdly dense'. Much of this network in rural areas, however, operated at a loss. The 1963 Beeching Report 'The Reshaping of British Railways' showed that one third of the rail network carried only 1 per cent of the freight and passenger traffic, and that one half of the mileage carried only 4–5 per cent of the total traffic. Beeching's drastic pruning of the network meant that

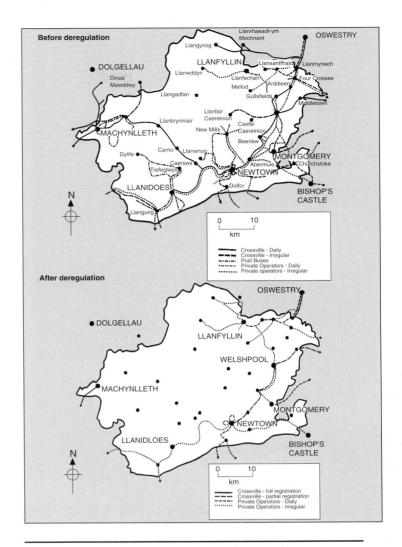

Figure 8.6.2   *Bus services in North Powys before and after deregulation*

by 1966 the route mileage had been reduced by 4500 miles and over 2000 stations closed, reducing the 1962 figure by 44 per cent (see Figure 8.6.3). Rural rail networks are now again under threat as a result of the privatisation of British Rail. However, current local initiatives indicate that rural rail services can play a vital part in rural life, given the right backing, and the appropriate marketing. In the south west the Devon and Cornwall Rail Partnership, involving a number of different groups, has focused its attention on two lines, the Exeter to Barnstaple line, 'The Tarka Line' (Figure 8.6.4) and the Plymouth to Gunnislake Line, 'the Tamar Valley Line'. A 7 per cent increase in passenger traffic has resulted from a number of initiatives aimed at the leisure market. Similar developments

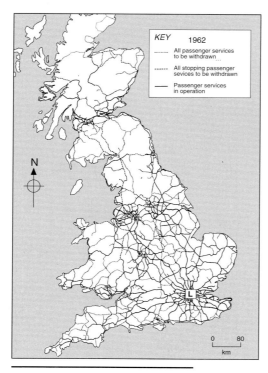

Figure 8.6.3   *The Beeching proposals*

Figure 8.6.4   *The Tarka Line at Copplestone, Devon*

Weymouth railway in Wessex, and the Penistone line in Yorkshire. Given the right co-operation between different groups, it is possible that rural railways can become a focus for sustainable development in the countryside.

STUDENT   **8.7**   ACTIVITY

1 Summarise the main problems facing those that rely on public transport in rural areas.
2 What arguments could be sustained for the operation of an integrated rural transport system, involving the public, private and voluntary sectors?

are occurring in other parts of the country, including the long-standing promotion of the Settle to Carlisle railway for both leisure and business purposes, and partnership plans for the Westbury–

ESSAYS

1 Explain why management issues vary so widely, in both nature and scale, across the rural regions of Britain and Ireland.
2 In the future, lifestyles in rural areas will show increasing divergence. Discuss this statement in relation to different social groups in the countryside.

FIELDWORK OPPORTUNITIES & PROJECT SUGGESTIONS

1 Select two or three villages which are reasonably close to one another. Ascertain from the local authority (district council) the different types of housing tenure in the villages (these are likely to be private owner occupied, public sector for rent, and private sector for rent). In particular assess the importance of housing associations in providing houses for rent. Carry out a housing quality, and environmental quality survey in the different areas of tenure, and make the relevant comparisons between them.

2 Select a number of villages (up to fifteen) in a rural area. For each village make a list of the services that it provides. Use a directory from the local library to determine the level of service provision in the villages in the past. Obtain population figures (1991 census) for each of the villages, and, if possible obtain figures for past censuses. It should be possible to draw a series of simple graphs to show how the level of services has changed over recent years, and how population levels have changed.

Managing Transport Systems

... transport should be made available to meet the needs of the population and ... everybody could expect a minimum level of mobility, almost as a right

All transport should be provided by the private sector, services should be determined competitively, not in a co-ordinated fashion, and fares should be market priced.

Two views from David Banister: Transport in Policy and Change in Thatcher's Britain, edited by Paul Cloke.

KEY IDEAS

- in Britain the management of transport services has largely passed into the private sector: in Ireland it has remained in the public sector
- the private motor car has increasingly come to dominate transport management issues
- the provision of cheap and efficient public transport is seen by many as a key management objective
- the effect of transport systems on the environment is now a national issue in both Britain and Ireland
- management of air and noise pollution requires effective policies that will create a more healthy environment

9.1 Introduction

Management of the transport environment in Britain has undergone far-reaching changes in the 1980s and the first half of the 1990s. After the election of the Labour Government in 1945, almost all of the transport services were nationalised. In the 1980s and 1990s the reverse occurred, with most of the transport industries being returned to the private sector. Motives for privatisation have been partly ideological, and partly financial. Ideologically, the private sector is seen as having the vital commercial stimulus of competition. From a financial point of view, the privatisation of transport sees the Government gaining large sums from sales of companies, and also having to pay less in subsidies and the servicing of capital debts in the public sector.

Apart from nationalisation a number of other important management changes have occurred:

- the planning framework of transport policy has been dismantled, so that the Government now deals directly with districts and other transport bodies, such as London Transport;
- bus services have been deregulated (see Chapter 8, pages 153), with commercial operators running a basic service, complemented by subsidised services where socially essential;
- an important number of safety measures have been introduced;
- traffic in towns has been identified as a key issue, and a number of controlling measures introduced;
- after the publication of the Royal Commission Report on Transport and the Environment, focus shifted sharply on to this new issue. Unleaded petrol was taxed at lower levels and other measures were introduced under the Environmental Protection Act of 1990.

In the mid 1990s discussion began to centre on a greater emphasis on the environment, more priority for public transport, and less dependence on the car. Although public support for these ideas is clear, there is, as yet, no sign of an obvious policy shift.

Transport policy in Northern Ireland is essentially roads-based, with much of the original rail network closed. Ulsterbus is responsible for all inter-urban and rural bus services. In the Republic of Ireland there is still a strong level of state involvement in the management of transport systems.

9.2 Case Studies in Managing Transport Systems

With the enormous increases in vehicular traffic in the second half of the twentieth century, the management of road systems has become a much more complex task. With the completion of the M40 much of the basic motorway network in Britain is in place, and the government's stated priority is now to 'make more use of our existing roads', with 'selective improvements through new construction, such as providing much needed bypasses and removing bottlenecks'. The first case study looks at road management in the late 1990s, with brief examinations of two by-

pass issues, at Newbury, and Melbury Abbas in Dorset. The privatisation of British Rail has attracted much media attention in the 1990s, and a review of the main management implications follows. A brief contrast with the Irish Railway system complements this study. The question of a fifth terminal at Heathrow is currently subject to a Public Inquiry, and the main issues involved are examined in a further case study. The environmental impact of the growth in all forms of transport is studied in a final section.

9.3 Managing Britain's Road System

The 1980s have been hailed by many as the decade of the car. Thatcher's 'great car economy' saw traffic increasing by 40 per cent and the numbers of cars and taxis by 30 per cent. Over two-thirds of households had at least one car, and found its use more or less indispensable (see Figure 9.3.1). Cars are used for 70 per cent of all journeys in Britain (see Figure 9.3.2): public transport can only offer a real alternative in congested urban areas or over very long distances. The 1990s and the early years of the next century clearly present a considerable challenge to those responsible for managing road systems and their traffic (Figure 9.3.3).

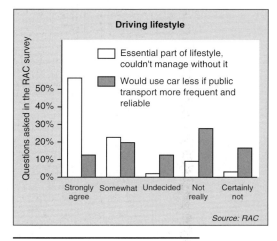

Figure 9.3.1 *How important is a car?*

Proportion of journeys (%)	Proportion of distance travelled (%)		Proportion (%) of journeys for each purpose made wholly or mainly							
			on foot	by car			by local bus	by rail	in other ways	
				Driver	Passenger					
29	40	leisure	12	37	74	37	6	2	6	100
23	15	other personal business	7	52	83	31	6	1	3	100
20	20	to and from work	5	58	72	14	10	5	8	100
19	11	shopping	10	43	70	27	16	1	3	100
5	11	business	3	81	92	11	3	3	3	100
4	2	education	18	6	38	32	21	3	10	100
100	100	all purposes	9	46	74	28	9	2	6	100

Figure 9.3.2 *Reasons for travel*

STUDENT   9.1   ACTIVITY

Read the extract from the *Independent*, 'End of the road for an impossible Tory dream' (Figure 9.3.5).
1 Why was Budget Day, 1995 'the worst day for Britain's infrastructure since the Romans left'?
2 To what extent do you agree with this description of the Budget decisions.
3 What 'difficult choices' now face Government transport managers?

When the situation described in the article is compared with the road improvement programme of the last 25 years, the change in policy becomes starkly clear. Since 1970 the length of the motorway network has been trebled (Figure 9.3.4) and since 1979 more than £24 billion has been spent on upgrading motorways and trunk roads (see map, Figure 9.3.6) and more recently, in the period since 1985/6 the Government has supported an £8.7 billion investment in the local road network. Figure 9.3.7 shows the proposed Government expenditure on transport for the closing years of the 1990s. The most recent review of the trunk road construction programme was in November 1995. In that review, each scheme was examined on a variety of grounds, and 77 were withdrawn either because they were environmentally unacceptable or because they were unlikely to be built in the foreseeable future. The Government has been keen to involve private sector finance in important new

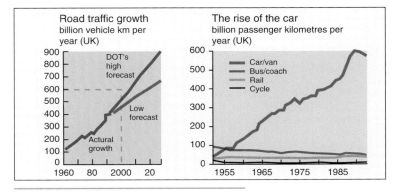

Figure 9.3.3 *Increase in road traffic since 1960*

road projects. However, the projects were essentially ones over which the private sector would have monopoly control, which included:
● the Queen Elizabeth II Bridge at Dartford
● the Second Severn Bridge
● the proposed Birmingham Northern Relief Road (see Figure 9.3.8)

Figure 9.3.4 *The M5 near Exeter*

End of the road for an impossible Tory dream

ONE of the great planks of Tory dogma was quietly, almost secretly buried last week when Kenneth Clarke and the Transport Secretary, Sir George Young, combined to ditch the Government's previously much-cherished road-building programme. The years of telling us that roads are essential for economic prosperity are now history. The national road-building programme will now consist of a few dribs and drabs, the odd bypass or trunk road widening scheme.

It is a momentous event, yet it was deliberately hidden among all the other news of the Budget because it begs more questions than the Government at present is able to answer. Indeed, there was more than a touch of dishonesty about what happened on Budget day. The Chancellor, in his characteristic bluff way, spoke of an extra £500m for roads under the Private Finance Initiative.

This is not extra money for roads – nor, as Sir George tried to depict it, just another way of bringing about the same level of roadbuilding. In fact, there has been a massive drop in the annual expenditure earmarked for national road schemes, from its peak of £2bn last year to £1.5bn, and we learnt in the Budget that it is to go on falling. But the more lasting effect, again revealed on Tuesday, is that under a review of the programme 117 schemes, some 60 per cent of the total, have either been permanently abandoned or put on hold. As a result, barely a handful of schemes will be started between now and the general election.

The roads lobby was appalled at both the decision and the way in which it was disguised, calling it the "worst day for Britain's infrastructure since the Romans left". They are also sceptical of the Private Finance Initiative's ability to deliver any roads quickly.

This is the sad end of the whole vision behind the roads programme, which was first set out in a rather thin White Paper called *Roads to Prosperity* in 1989. The gist of the argument then was that Britain needed "a major expansion of the Government's programme for building and improving inter-urban roads" to "meet the forecast needs of traffic into the next century". These were heady times for the roadbuilding industry, as it seemed that the Government genuinely believed it could build itself out of the traffic congestion crisis.

The problem was that there was never any hope of doing so. Traffic was expected to rise by between 142 per cent and 834 per cent between 1988 and 2025, and there was never any chance of increasing the capacity of Britain's roads by that amount. Money was being pumped into a programme that at best stopped things getting worse quicker. Finally, the Treasury said no more.

But why didn't Sir George proudly boast about his new policy, rather than slipping it through as part of the Budget? Because in representing such a massive U-turn it was simply too embarrassing, and to boot because he is offering nothing in its stead.

He made a few token comments about making "more efficient use of the roads we have", but this would cost a lot of money too.

In Britain, we are waiting for the private sector to develop the infrastructure and pay for its installation, because the Government refuses to put in any seed-corn funding. Japan is already using technology to reduce congestion, while in Britain we are years away from even starting pilot schemes. This will be a great missed opportunity, since Japanese equipment manufacturers will be in a position to flood our market in the same way they have done with cameras and Walkmans.

The destruction of the roadbuilding programme in the Budget signifies that transport policy has been taken over by the Treasury. If Sir George wants to retain his credibility, he needs to wrest back the initiative. Earlier this year, Dr Mawhinney launched a transport debate which seemed genuinely to be asking the right questions about transport policy.

When Sir George publishes the results of the debate, which he has said he will do early next year, he must do more than reiterate platitudes about congestion and actually suggest radical ways of tackling the crisis. He must actually begin to make difficult choices which will antagonise a lot of people, for example by restricting car parking in towns, reducing speed limits or turning over road space to cyclists.

Now that few new roads are to be built, it is only through such courageous measures that the inexorable clogging up of our roads can be halted. There is no shortage of examples from abroad where all sorts of well-tried schemes, ranging from measures to increase bus usage to building light rail systems, are being implemented. Some of these are cheap, others cost a lot of money. Sir George must now let the Treasury have its way by stopping roadbuilding, but in return he has to persuade them to cough up for alternative transport policies that will stop the steady drift towards gridlock.

The Independent 2 December 1995

Figure 9.3.5   End of the road for an impossible Tory dream

Tolls charged on each of these projects would contribute towards the building and operating costs.

In its document 'Transport: the Way Ahead', the Government emphasises the importance of bypasses (Figure 9.3.9) in its future programme. It sees their principal advantages as:

- taking heavy through traffic out of town centres and villages, and thus preserving their historic identity;
- making towns and villages cleaner, safer, and healthier places to live and work;
- using less space for carriageways in towns so that there is:
  - more pedestrian-friendly town centres
  - better access for local people
  - opportunity for environmental improvements
  - more space for trees;
- relieving bottlenecks, so bringing benefits to through traffic.

The state of bypass construction in the mid 1990s is shown in Figure 9.3.10.

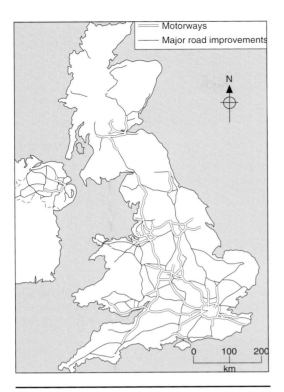

Figure 9.3.6 *The British motorway network in 1994*

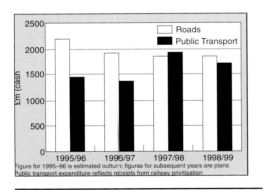

Figure 9.3.7 *Government expenditure on transport*

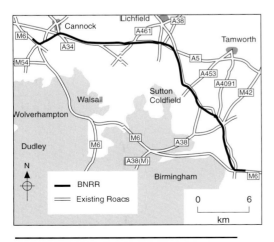

Figure 9.3.8 *Birmingham northern relief road*

Figure 9.3.9 *The Bridport Bypass, Dorset*

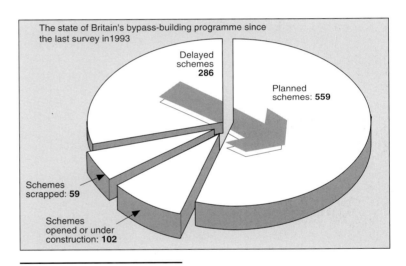

Figure 9.3.10 *Bypassing Britain*

THE NEWBURY BYPASS

The Newbury bypass commanded much media attention in 1996, largely because of the activities of well-organised protesters, who sought to disrupt preparatory work along the route. The final approval of the bypass caused some considerable surprise, since the Transport Secretary had deferred a decision on the scheme a year earlier, in a newly created atmosphere of open debate on major transport issues. Argument has centred around both the need for a bypass, and the choice of the western alternative as the selected route (Figure 9.3.11).

'Mawhinney's decision shows the whole transport debate was a sham. It was a terrible betrayal.'

Roger Higman, Roads Campaigner of Friends of the Earth on hearing that the Transport Secretary, Brian Mawhinney had approved the proposed Newbury bypass.

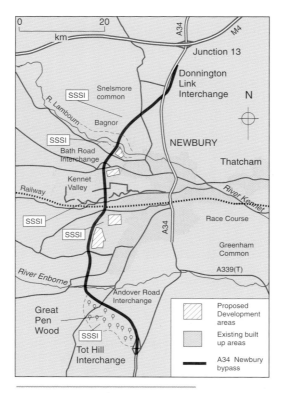

Figure 9.3.11   The Newbury bypass

The case for a Newbury bypass rests on the following points:
- the A34 through Newbury is part of Euroroute EO5 which runs 1920 km from Glasgow to Burgos in northern Spain. It is all dual carriageway, with the exception of 5 km to the south of Newbury;
- pollution levels along the A34 show a 300 per cent increase in NOx gases compared with the rest of Newbury;
- increased numbers of cases of asthma have been recorded amongst young children living near the A34;
- noise levels near the A34 have become unacceptably high;
- traffic flows along the A34 average 50 600 vehicles per day;
- development pressures in the Newbury area are likely to increase traffic volume considerably.

The western route was selected as the preferred route over alternative routes through the centre of Newbury, and to the east of Newbury. Central routes were rejected because of demolition of houses, disruption of Newbury's local traffic during construction, the physical dominance of flyovers and slip roads, their failure to reduce levels of noise and pollution within the town, and the cost (twice as much as a

western route). Eastern routes were rejected because of damage to a nature reserve, greater noise levels for more homes than a western route, the need for a large A34/A4 interchange, tight bends around the racecourse, a longer route, and cost (70 per cent higher than the western bypass).

STUDENT   9.2   ACTIVITY

1 Discuss some of the environmental issues that are likely to cause controversy during the discussion of alternative routes for a bypass.
2 Use the map of the selected bypass route (Figure 9.3.11) to identify some of the contentious points on the western bypass likely to have led to a conflict of views.
3 Draw up a model of a cost benefit analysis suitable for assessing the general impact of bypasses.

THE C13 IMPROVEMENT AND THE MELBURY ABBAS BYPASS

It is perhaps unusual for a major road improvement scheme to be proposed for a C Grade road, since such roads in the country usually carry relatively little traffic apart from that which is generated locally. The C13, together with the A350 is one of a number of corridors that lead northwards from the developing cross Channel ferry port of Poole in Dorset (see Figure 9.3.12). None of these corridors is adequate to carry forecast levels of traffic from Poole in the future, but the A350/C13 corridor was selected for improvement because the local

Figure 9.3.12   Corridors leading north from Poole

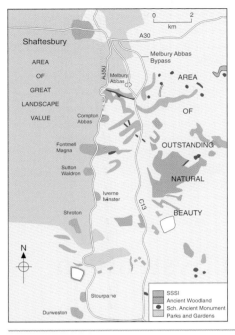

Figure 9.3.13   *The C13 improvements and the Melbury Abbas bypass*

traffic along this corridor could not be reasonably expected to divert to any of the other corridors. Routes along the line of the A350, which links a series of villages at the foot of the Dorset Downs were examined and rejected on the grounds that they did not solve the environmental problems of these settlements. The A350/C13 corridor is shown on the larger scale map (Figure 9.3.13).

The new road along the course of the C13

Figure 9.3.15   *Countryside through which the Melbury Abbas Bypass is routed*

would provide a modern, high quality route for long distance traffic, including heavy goods vehicles from Poole to the main road network of the rest of the United Kingdom.

The main environmental benefits of the scheme are seen as follows:

- in 2015, 727 properties would experience noise levels less than today;
- in 2015, 412 properties would experience an improvement in air quality;
- **severance**, and general stress among inhabitants of the villages would decrease as a result of lower traffic flows;
- more vulnerable road users, e.g. cyclists would be safer;
- the adverse effect of traffic on listed buildings, registered parks and conservation areas would be reduced;
- the number of collisions of vehicles with buildings would be reduced;
- in Shaftesbury, 49 buildings would experience noise reductions from today's levels;
- in Shaftesbury, 16 residential properties would experience better air quality.

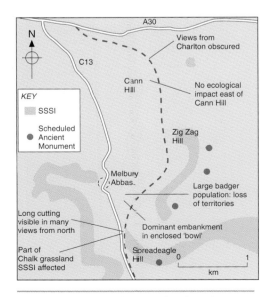

Figure 9.3.14   *Landscape and ecological impacts, Melbury Abbas bypass*

Figure 9.3.16   *The existing C13 route through Melbury Abbas*

'Britain's railways are on the threshold of a renaissance, achieved through . . . new private sector ideas and established railway industry expertise.'

Sir George Young, Transport Secretary, writing in the *Independent on Sunday*, May, 1996.

STUDENT   **9.3**   ACTIVITY

Look at Figures 9.3.13, 9.3.14, 9.3.15 and 9.3.16.
1 What are the main likely adverse environmental impacts of the C13 and Melbury Abbas improvement?
   List these impacts under four headings
   a landscape
   b ecology
   c archaeology
   d farming and land use

2 What additional impacts would the scheme have on the areas adjacent to the two roundabouts at either end of the C13?
3 The C13 is already being called 'Dorset's Newbury bypass'. Identify any similarities between the two schemes. Are there any significant differences?
4 Bearing in mind the Government proposals for road programmes in the next few years, do you think that this scheme should go ahead?

'If you believe . . . that Britain's railways need a renaissance for all kinds of good reasons, social, environmental, economic, then their hurried ramshackle privatisation is probably the largest and certainly the most avoidable disaster to be visited on this country during its Major years.'

Ian Jack, writing in *The Great Train Robbery, Independent on Sunday*, May, 1996.

9.4 Managing Railway Systems in Britain and Ireland

THE PRIVATISATION OF BRITISH RAILWAYS

The privatisation of British Railways has certainly been one of the most controversial of all of the moves from the public to the private sector within the last 15 years. Britain's railways were nationalised by the Labour Government in 1948, and have remained in the public sector until 1996. The changes resulting from the privatisation will be the most fundamental since the Beeching Report of 1963 (see Chapter 8, pages 153–4). Figure 9.4.1 shows the railway network at the time of privatisation in the mid 1990s. Privatisation, however, is fundamentally different from the Beeching reforms. Whereas Beeching was principally concerned with the pruning of a network, within which many lines were unprofitable, privatisation is concerned with the reorganisation of the manner in which Britain's railways are run. It involves a change from public to private sector finance, and total restructuring of the way in which the railways are managed. The basic structure of the railways of Britain after privatisation is shown in the diagram below (Figure 9.4.2). Trains on the railways will be run by 25 Train Operating Companies (TOCs) (Figure 9.4.3). Companies wishing to run the TOCs will have to bid competitively

for the **franchise** to operate the trains over the routes allocated. The rolling stock will be owned by three companies known as ROSCOs (Rolling Stock Companies). These lease rolling stock to the TOCs. Railtrack owns the track and infrastructure

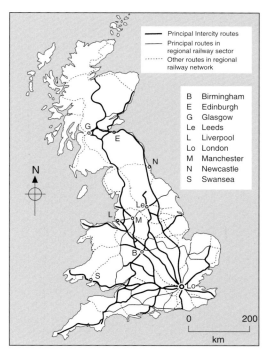

Figure 9.4.1   *British rail network at the time of privatisation*

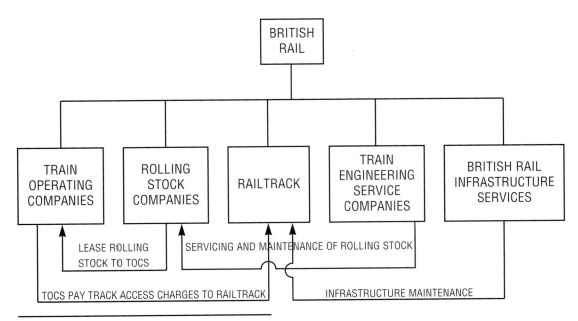

Figure 9.4.2   Structure of railways after privatisation

(stations and signalling) over which the TOCs run their services. The TOCs pay track access charges to Railtrack in order to run their services over the routes for which they are responsible. Engineering Services are provided by Train Engineering Service Companies (TESCOs). All infrastructure maintenance is carried out by British Rail Infrastructure Services (BRIS).

Regardless of the ideological issues involved in privatisation, it is important to review the advantages and disadvantages of privatisation of Britain's railways. The advantages may be summarised as follows:

- restructuring frees the railways from the central bureaucratic control;
- the creation of smaller companies will enable managers to serve local markets better;

- railways, with greater commercial incentives in operating through the franchise scheme, should be able to compete more effectively for a greater share of the transport market;
- a commercially adroit service should be able to meet the customers' needs more closely, and offer a better service;
- railways will now have access to private capital, rather than from the Treasury, where there were always other competing claims from other programmes such as health and education. Private capital will be available for major investments that are badly needed for a fully modernised railway.

The disadvantages of rail privatisation would appear to be:

- a privatised rail service will seek to encourage only the most profitable routes;
- closure of some of the socially necessary services in remote rural areas is possible;
- frequency of trains may be reduced by only running those that are likely to be heavily used;
- with so many different companies operating, integration between them will be difficult, since, in some cases they will be competing against one another;
- railways run solely for profit may not make the necessary investments in infrastructure and rolling stock that are badly needed;

Figure 9.4.3   Weymouth to London train, South West Trains at Wareham Station, Dorset

- there is a concern that fares may have to increase in order to pay for the necessary essential investment.

What are the geographical implications of this fundamental change in the management of Britain's railways? The most obvious is a possible reduction in the network. Although there have been closures since the Beeching cuts of the 1960s, much of the trunk network has remained intact. It has been suggested that the first casualties in any post-privatisation pruning of the network could be branch lines in the south west, the north west, East Anglia, and the north east. Many of these lines, although lightly used, are regarded as socially necessary, and are important in the promotion of tourism in the respective regions. The possibility of higher fares, and frustration in making journeys that involve more than one TOC may lead to passengers turning to their cars as alternatives, thus increasing pressure on roads. If, on the other hand, privatisation is as successful as some hope, then there could be a resurgence of the railways. Both passengers and freight could be persuaded to return to the railways and there would be less traffic on the roads, and lower levels of environmental pollution, both in towns and rural areas.

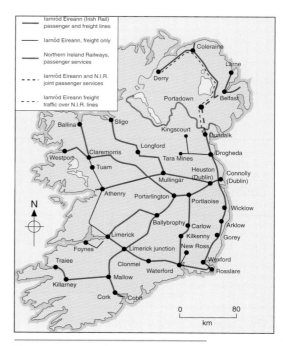

Figure 9.4.4 *Ireland's railway network*

STUDENT   **9.4**   ACTIVITY

1 Some people have described rail privatisation as a 'poll tax on wheels' (the poll tax was one of the most unpopular measures introduced by the Thatcher Government). What aspects of the new management of British Railways are likely to make it unpopular?
2 Why is it virtually impossible to renationalise the railways in the future?
3 What, in your opinion, are the factors most likely to make rail privatisation successful?

IARNROD EIREANN: IRISH RAIL

The Irish Transport Authority (Coras Iompair Eireann) was formed in 1945, and it took over the assets of the Great Southern Railway, which owned most of the trunk lines in the Irish Republic. Coras Iompair Eireann (CIE) was nationalised in 1950, and in 1958 it took over the lines

within the Republic of the Great Northern Railway, which, until then had remained independent. In 1987 CIE was split into three subsidiary companies dealing separately with the railways, the city bus services and non-city bus services. Iarnrod Eireann is responsible for the railways. The main features of the Irish rail network (Figure 9.4.4) are:

- Dublin acts as a central focus for lines that radiate out to the west and south coasts;
- Dublin acts as a centre for a dense network of suburban lines (DART, see Chapter 6, page 112);
- a limited number of cross country lines, although some of these carry only freight.

In 1984, the Irish Government took the decision that there should be no further substantial development in railways. In 1987, however, **ERDF** money became available, but most of this was assigned to developments in the Dublin suburban area, with little available for the important intercity routes between Dublin, and other major cities such as Cork, Limerick and Sligo (see Figure 9.4.5). A further change in policy resulted from intense public pressure, following alarming falls in the standard of service on some Intercity lines.

Investment in all of the rail system is now official Government policy. Passenger miles on the Intercity have now shown an important increase, although freight-tonnes carried have dropped since a peak in the early 1980s.

Figure 9.4.5   Intercity train, Ireland

STUDENT   9.5   ACTIVITY

1 What are the disadvantages of operating a railway system in a small country such as the Irish Republic?
2 Is public sector control of railways more appropriate in a small country?

9.5   Heathrow: Terminal 5?

Proposals for airport expansion are always amongst the most contentious. The Terminal 5 scheme at Heathrow is currently the subject of a Public Inquiry that will last at least 18 months and cost more than £10 million. Heathrow is the world's busiest airport, handling more than 51 million passengers a year, and over 408 000 air transport movements (atms). Demand is set to grow at a rate of 3.9 per cent per annum to 2006, and 3.4 per cent per annum to 2016. Figure 9.5.1 shows the relationship between capacity and demand

Figure 9.5.2   Site for Terminal 5, Heathrow

at Heathrow to the year 2016. Terminal 5 will increase the capacity from 50 million passengers per annum (mppa) to 80 mppa. The map (Figure 9.5.3) shows the location of the new Terminal 5 in relation to the existing facilities at Heathrow. If approval is given, it will be built at Perry Oaks on a 250 hectare site, technically within the Green Belt, currently occupied by a sewage sludge works. The site lies within the existing airport perimeter, between the two runways. A spur road would give access to the M25 to the west. Fifteen houses would have to be demolished, although nine are owned by British Airports Authority (BAA). Within the site 80 observed species of birds have been recorded, of which ten are currently under protection.

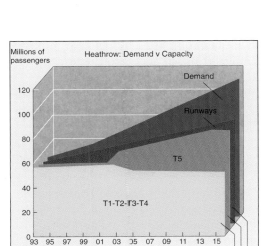

Figure 9.5.1   Demand and capacity at Heathrow

'... without Terminal 5 the cost to British commerce and industry would increase by a figure approaching £1 billion a year.'

Sir Colin Marshall, Chairman of British Airways, writing in *The Times*, 16 May 1995.

'We have just had air pollution alerts on the M25 and the Department of Transport has announced plans for speed restrictions. The whole area is polluted and congested without Terminal 5, on its own bigger than Paris's Charles de Gaulle or Amsterdam's Schipol airport.'

Dermot Cox, Chairman of the Heathrow Association for the Control of Aircraft Noise (HACAN).

Two rivers within the affected area would have to be re-routed.

BAA maintain that Terminal 5 is essential if Heathrow is to remain competitive internationally. Major continental airports, such as Paris, Charles de Gaulle, Amsterdam, and Frankfurt have already considerable plans for expansion. If Heathrow does not provide the extra capacity, then airlines will choose to use other European airports, with losses of up to £600 million in export earnings by 2005. Seventy-eight thousand people are employed directly or indirectly by the airport, with a **multiplier effect** extending to over 190 000 jobs. Terminal 5 will protect 6000 jobs, and is likely to create 8000 new ones, as well as 3000 temporary new construction jobs. The new terminal would have connections to the new Heathrow Express rail link to central London, and to the Piccadilly Line, thus encouraging the use of public transport (about 40 per cent of people using Terminal 5 are likely to travel by public transport).

Opposition to Terminal 5 is co-ordinated by LAHT5 (Local Authorities Heathrow Terminal 5 Group). Their case against the building of Terminal 5 rests on the following points:

- increased capacity at Heathrow (with Terminal 5) is an underestimate: it could exceed 100 mppa;
- alternatives, such as Stansted could effectively handle some of the increased capacity envisaged;
- proposals for Terminal 5 run contrary to development plans and in particular to Green Belt policies;
- high levels of employment growth at Heathrow (with Terminal 5) would increase development pressures;
- additional traffic generated by the new proposals will have an unacceptable impact on the congested road network of the area;
- air noise is likely to increase, despite some improvements in aircraft noise;
- air quality is likely to deteriorate, and particular concern is felt over concentrations of benzene (a carcinogen) and particulate matter;
- damage to and unbalancing of local ecosystems is likely to occur.

BAA have sought to reassure local people that:

- no third runway at Heathrow is necessary: it is terminal capacity that needs to be increased, not runway capacity;
- Terminal 5 will not call for an increase in the quota of night flights;
- Terminal 5 will not require a 14 lane M25 (this expansion has now been ruled out anyway);
- BAA is aiming for a 50 per cent public transport use in access to the airport, thus reducing traffic congestion;
- the overall noise levels will be no worse when Terminal 5 is fully operational than it is today.

STUDENT   **9.6**   ACTIVITY

1 Should national economic benefits of such a proposal as Terminal 5 outweigh local considerations of a deteriorating environment?
2 Why do you think the most recent poll showed nearly 50 per cent of local people in favour of Terminal 5?

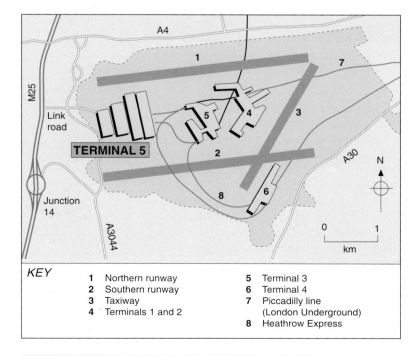

Figure 9.5.3   *Terminal 5 in relation to existing airport facilities*

KEY

1 Northern runway
2 Southern runway
3 Taxiway
4 Terminals 1 and 2
5 Terminal 3
6 Terminal 4
7 Piccadilly line (London Underground)
8 Heathrow Express

9.6 Transport and the Environment

6 May 1995 was a fine late spring day. Yet *The Times* reported that Britain was suffering from 'one of the worst episodes of air pollution since the 1950s as large areas of the country are blanketed by smog formed by a cocktail of toxic gases'. Cities across the country were affected, and those suffering from lung disorders were advised to stay indoors and avoid strenuous exercise. Rural areas did not escape, and much of the Peak District and the North Yorkshire Moors were covered in a haze caused not only by traffic fumes, but also power station emissions. Levels of nitrogen dioxide reached critical health levels by mid-morning in places such as Sheffield, and suburban and central London. Ozone levels breached international guidelines. The day before, the *Daily Mirror* carried banner headlines 'SMOG WARNING – Forecast grim for Britain's 3 million asthma sufferers', and complained that 'millions of people woke up yesterday (4 May 1995)

with streaming eyes, headaches, breathing trouble and a choking cough.'

Although pollution levels on one day can reach alarming levels, a more general survey also gives rise for concern. Figure 9.6.1 shows concentration levels for ozone, and nitrogen dioxide. Ozone is not emitted directly by motor vehicles, but arises from a series of complicated chemical reactions, partly driven by the sun, from other gases, particularly nitrogen dioxide and hydrocarbons, such as benzene. Surprisingly some of the highest concentrations of ozone are in the uplands of Britain, the result of complicated chemical and meteorological processes. Over half of the nitrogen dioxide present in the atmosphere is derived from the exhausts of motor vehicles. Figure 9.6.2 shows the present distribution of asthma cases, and the rate of increase of asthma in Britain.

'At present pollutants from vehicles are the prime cause of poor air quality that damages human health, plants, and the fabric of buildings. Noise from vehicles and aircraft is a major source of stress and dissatisfaction, notably in towns, but now intruding into many formerly tranquil areas.'

Concluding Chapter: Royal Commission on Environmental Pollution, October 1994.

Ozone

Parts per Billion
- 34 and above
- 32 – 33.99
- 30 – 31.99
- 28 – 29.99
- Below 27.99

Nitrogen Dioxide

Parts per Billion
- 20 and above
- 16 – 19.99
- 12 – 15.99
- 8 – 11.99
- 4 – 7.99
- Below 4

0   50
km

Figure 9.6.1 *Concentration levels of ozone and nitrogen dioxide levels in Britain*

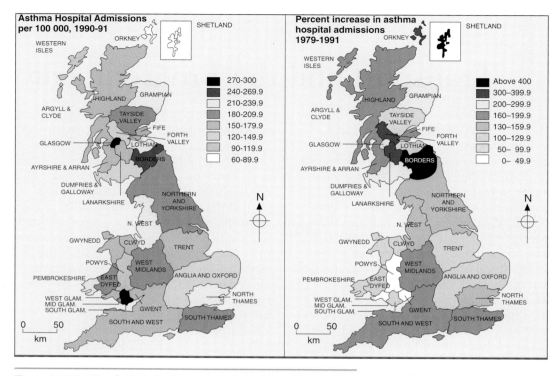

Figure 9.6.2   *Distribution of asthma cases and rate of increase of asthma*

STUDENT   **9.7**   ACTIVITY

1 Compare the distribution of concentrations of ozone and nitrogen dioxide in Britain.
2 What would appear to be the main factors responsible for these distributions and the differences between them?
3 Comment on the incidence of asthma, and its rate of increase as shown on the maps in Figure 9.6.2.
4 How might they be related to the distribution of the pollutants shown in Figure 9.6.1?

In January 1995, the Department of the Environment, in a management response to the Royal Commission, launched the Paper 'Air Quality: Meeting the Challenge'. Local councils are now responsible for drawing up plans for air quality management areas within their districts. Health standards for nine pollutants are to be announced, with the first three being benzene, 1.3 butadiene (another carcinogen) and carbon dioxide. A 'dirty drivers' campaign has been launched by the Department of Transport, although some feel it will not have much overall effect, since it does not introduce any new air quality standards. EU directives on air quality are likely to be tough, with a strict timetable for achieving them. In July 1996 the National Air Quality Strategy was announced. Maximum concentrations for eight known pollutants were set. The enforcement of these controls on pollution is to be in the hands of local authorities, although resources to fund this strategy are likely to be in short supply.

ESSAYS

1 What are the difficulties in achieving a sustainable transport policy?
2 To what extent are transport policies a reflection of Britain being small and densely populated?

Managing Leisure and Tourism

KEY IDEAS

- Leisure and tourism are both fast growing industries in Britain and Ireland, largely as a result of greater amounts of leisure time, and more disposable income
- The provision of leisure and tourism services involves both the public and the private sector
- Leisure management must provide a range of services which respond to the varied demand created by the population
- Management of tourism has to strike a balance between service provision and the conservation of both the built and the natural environment

10.1 Introduction

By the late 1970s Britain may be considered to have . . . (become) . . . one of the leading tourist destinations, with a highly developed tourist industry.

Burkhart and Medlik. Tourism.

Leisure and tourism are both increasingly important industries in Britain and Ireland. Britain's tourist industry now employs 1.5 million people and is responsible for 5 per cent of export earnings. Other details are shown in Figure 10.1.1. Management operates on a variety of scales from Government involvement at the highest level down to local management at town or even village scale. The private, public and voluntary sectors are all involved in managing tourism at the various levels. The structure of management of the tourist industry in Britain is shown in Figure 10.1.2. Within each country tourism is managed at an intermediate level by the Regional Tourist Boards. Funding for the British Tourist Authority, and the individual tourist boards comes from central government, which has further management functions in issuing broad guidelines, and encouraging various developments. In Northern Ireland, the Northern Ireland Tourist Board operates as

Volume of spending of tourists in the UK 1994

Millions	Trips	Nights	Spending
UK residents	109.8	417	£14 495
Overseas visitors	21.0	194	£9 919
Total	130.8	611	£24 414

International tourism receipts

1	USA	$56 501 mn
2	France	$23 410 mn
3	Italy	$20 521 mn
4	Spain	$19 425 mn
5	UK	$14 064 mn

Tourism in the UK economy

Economic indicator	£ billion 1994	Tourism's share
Gross domestic product	669	4.0%
Consumers' spending	428	6.3%
All exports	258	4.8%
Services exports	39	31.6%

Distribution of tourism 1994

Millions	UK residents		Overseas visitors	
	Trips	Spending	Trips	Spending
Cumbria	2.9	£425	0.30	£53
Northumbria	3.1	£355	0.44	£131
North West	8.6	£1 090	1.12	£395
Yorkshire & Humberside	9.3	£1 110	0.91	£220
Heart of England	9.9	£1 050	1.34	£378
East Midlands	7.6	£835	0.68	£188
East Anglia	10.2	£1 175	1.32	£390
London	8.6	£1 105	11.46	£5 281
West Country	15.1	£2 455	1.46	£399
Southern	10.6	£1 145	1.83	£590
South East	7.9	£900	2.03	£639
England	90.2	£11 650	18.07	£8 671
N.Ireland	1.2	£180	0.10	£54
Scotland	8.5	£1 310	1.77	£768
Wales	9.8	£1 075	0.69	£190
UK*	109.8	£14 495	21.03	£9 919

* UK includes Channel Islands and the Isle of Man.

Figure 10.1.1 *Britain's tourist industry*

one of four functional units within the Department of Economic Development, as shown in the diagram, Figure 7.3.3, page 124. During the ceasefire in Northern Ireland tourism showed a massive resurgence, with visitor numbers up 56% in the first six months of 1995.

In the Republic of Ireland tourism began to emerge as a significant element in the economy some 30 or 40 years ago. The Irish State body responsible for tourism is Bord Failte Eireann, which has much the same function as the British Tourist Authority. There are seven regional tourist organisations (one of which is operated by Shannon Development), which provide a

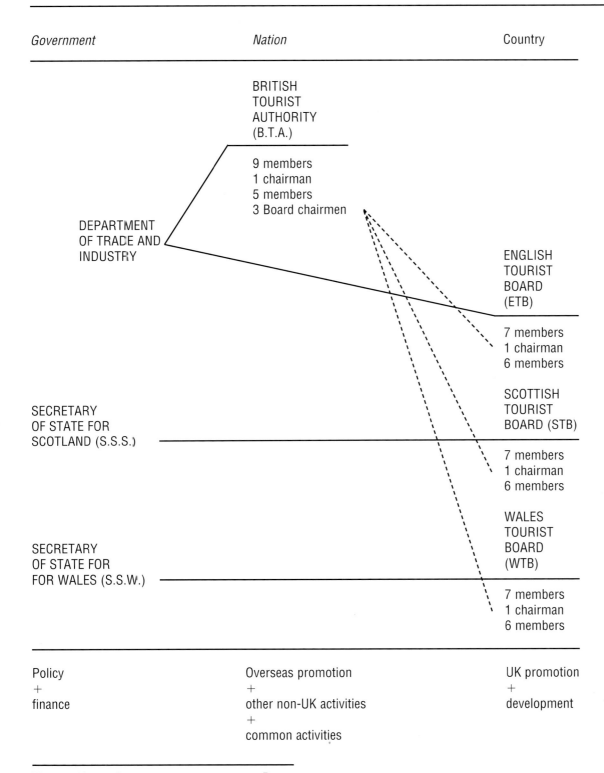

Government	*Nation*	Country
	BRITISH TOURIST AUTHORITY (B.T.A.)	
DEPARTMENT OF TRADE AND INDUSTRY	9 members 1 chairman 5 members 3 Board chairmen	ENGLISH TOURIST BOARD (ETB)
		7 members 1 chairman 6 members
SECRETARY OF STATE FOR SCOTLAND (S.S.S.)		SCOTTISH TOURIST BOARD (STB)
		7 members 1 chairman 6 members
SECRETARY OF STATE FOR FOR WALES (S.S.W.)		WALES TOURIST BOARD (WTB)
		7 members 1 chairman 6 members
Policy + finance	Overseas promotion + other non-UK activities + common activities	UK promotion + development

Figure 10.1.2 *Statutory tourist organisation in Britain*

network of information offices, and room-booking services throughout the country. Tourism has seen a steady growth in the Republic, with a 50 per cent increase in visitors from continental Europe between 1988 and 1994. A new operational programme for tourism was launched for a five year period in 1994, with an investment of £IR 652 million.

Within this broad framework of state-directed management, both public and private sectors provide important resources and services for tourism. For example, National Parks (supported by public funds) exercise important tourist management functions: the National Trust (supported mainly by donations and subscriptions) also plays an important role in this respect. In all cities and towns, local authorities are responsible for the management of tourism, although many of the facilities may be supplied and managed by the private sector.

In a similar way, both public and private sectors supply and manage leisure facilities. With the increasing demand for leisure provision, consequent upon more disposable income, and more available free time, there has been an enormous growth in the number of leisure centres, principally in urban areas. More recently the private sector has, however, increasingly used urban fringe or rural locations for the development of leisure parks.

10.2 Case Studies in the Management of Leisure and Tourism

The first case study reviews the management of one of the oldest of leisure facilities. Urban parks represent a well established tradition, founded often in the Victorian planning of towns and cities. A century later they still play a fundamental role in leisure provision in urban areas. The remaining case studies examine the management of tourism in a number of different locations, some urban, some rural. Over the last two decades, almost every town and village has sought to capitalise on its historic associations and traditions, and the brown direction indicators have become a familiar sight along the road network. Historic towns and cities, such as Bath and Chester in England, Stirling and Edinburgh in Scotland, and Armagh and Cashel in Ireland seek to offer the visitor a full and varied programme. Heavy tourist pressure needs skilful management in cities still adjusting to the demands of the motor vehicle. Salisbury, with its Cathedral and Close, its refurbished centre and quiet water meadows, and its nearby attractions of Old Sarum and Stonehenge is typical of the historic city promoting tourism, catering for visitors and managing traffic. Ancient sites, both accessible like Stonehenge, and more remote like the hill forts and stone circles in the uplands of

Britain have to be managed positively and skilfully. Hadrian's Wall is studied in the light of the new management plan recently introduced by English Heritage. Industrial archaeology has increasing appeal to the tourist, and the New Lanark settlement of Robert Owen, with its present population living and working in tenements and mills, is in many ways a unique attraction.

The scenic heritage of Britain and Ireland offers stunning variety to visitors from elsewhere in the islands or from overseas. Britain's longest National Trail, the South West Way, winding around the coast for over 900 km through four counties presents complex management issues that now will see a fully co-ordinated approach. AONBs whilst not having the complex planning mechanisms for recreation that exist in the National Parks, still need to be managed. The Malvern Hills, with their own elected Board of Conservators, still offer the ridge walks and wooded drives that so appealed to Sir Edward Elgar and his Edwardian contemporaries.

10.3 Managing the Urban Park: 'Leisure centres without a roof'

In most towns and cities there is a **hierarchy** of open spaces. In general, the larger parks will offer a wider range of facilities, and will attract visitors from a wider area than the smaller ones.

STUDENT   10.1   ACTIVITY

1 Study the map showing the distribution of open space on Portsea Island, Portsmouth (Figure 10.3.1). The survey, carried out in 1972, recorded the percentage of children in each park, the percentage of people that came on foot, and the number of visits that they made per week.
   a Use the bar graphs and pie segments to determine a possible hierarchy of open space in Portsmouth.
   b What other criteria could be used for determining a hierarchy of open space?

c What pressures would there have been on the open space in Portsmouth over the last 25 years? How would you have expected the use of these open spaces to have changed in that period? What management responses would be necessary?

The idea of hierarchy of parks is a useful management tool. In a review of open space available in the London Borough of Bromley, the management team decided that if use of the parks was to be maximised, a full audit, and new classification of the parks within the Borough would be necessary. They decided on a six fold classification of parks:
- ornamental town park/garden
- multi-purpose park
- natural park/woodland
- local open space
- sports grounds
- linear park/walk circular walk

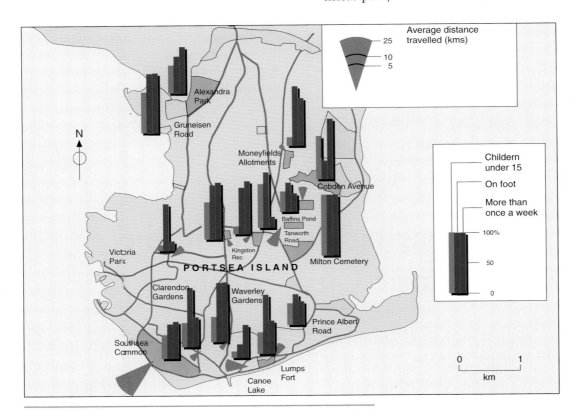

Figure 10.3.1   *Distribution and use of public open space on Portsea Island*

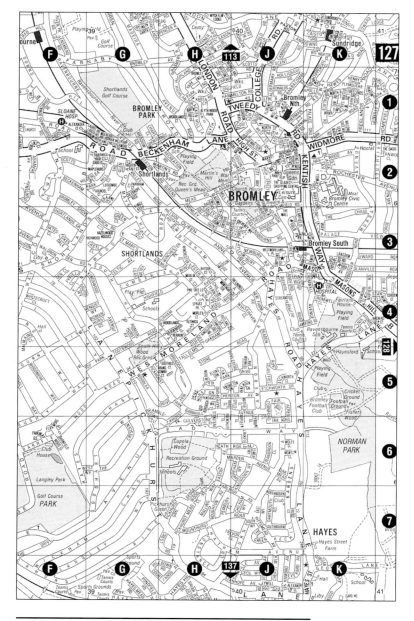

Figure 10.3.2 *Urban park and open space in part of Bromley*

1 Figure 10.3.2 shows part of the Borough of Bromley.
   a Trace off from the map all of the open spaces.
   b Using the evidence from the map, attempt to classify the different spaces according to that derived by Bromley Borough Council.
   c Draw up a table showing the different groups that might use the different open spaces.

Local authorities are now moving towards the idea of individual management plans for their parks. Such management plans might include:
- declared aims for the park;
- an appraisal of the park, including use, access, physical, ecological and visual analysis;
- an action plan with agreed priorities, areas for investment, and a development schedule;
- a design concept for the different areas of the park;
- a statement of how park management will operate;
- a statement of the role of partners, such as wildlife groups, sports clubs;
- an events programme, including opportunities for voluntary groups to run their own events;
- a framework for encouraging wider community participation.

'Salisbury is held by many people, and not all of them Wiltshiremen, to be the finest town in England.'

R. L. P. Jowitt: Salisbury.

10.4 Managing Tourism in the Historic City: Salisbury

The essence of Salisbury has nowhere been better captured than in Constable's well known painting of the Cathedral seen from across the water meadows of the River Avon (Figure 10.4.1). Within the City, the Cathedral (Figure 10.4.2), and Close are still the central attractions, but the whole of south Wiltshire, with its rich archaeological remains (Figure 10.4.3), and its clear Chalk streams, threading their way through broad open valleys, has much to offer the tourist.

Figure 10.4.1   *Constable's painting of Salisbury cathedral and the water meadows*

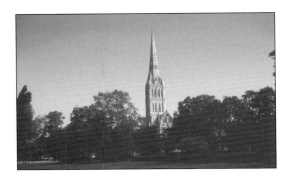

Figure 10.4.2   *Salisbury Cathedral*

As with many historic cities, the mediaeval fabric of the town is not well suited to the heavy vehicular traffic that converges on Salisbury from eight different directions. A long standing issue has been the battle fought over the promised bypass to replace the out-of-date ring road (Figure 10.4.4). A recent report, however, suggests that the

Figure 10.4.3   *Stonehenge, Salisbury Plain*

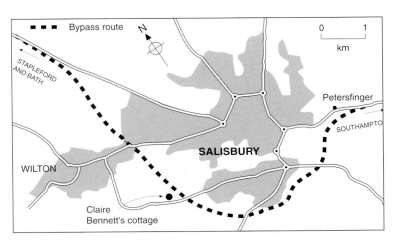

Figure 10.4.4   *The Salisbury bypass*

basic problem is not through traffic, but local traffic. Of the 50 000 daily traffic movements on the A36, only 3000 emerged from the far side of the city, suggesting that 94 per cent of the traffic is local or tourist traffic. Hence a major traffic management problem has inevitably emerged in the city centre. The 1993 Local Plan laments the fact that '. . . traffic

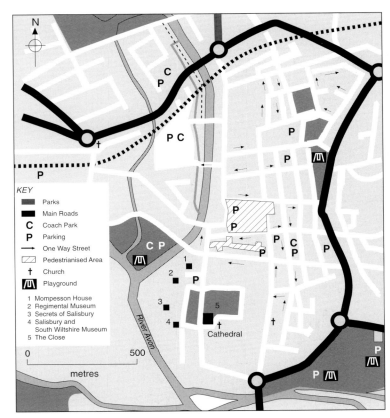

Figure 10.4.5   *Salisbury city centre, tourist attractions and traffic management*

restraint and pedestrianisation . . . have largely gone unimplemented, and as a consequence the city lags conspicuously behind equivalent historic towns, both home and abroad, where real progress has been made in limiting vehicular access . . .'. Figure 10.4.5 shows that present management is restricted to a one way system, and some pedestrianisation in central areas.

The map also shows the location of the principal car and coach parks. At present there are just over 4000 public car parking spaces available. A shortfall of some 2700 spaces is forecast for 2006, and this would be considerably greater on the two busy market days. Future management of parking requires more short term capacity, some of which will have to be provided in multi-storey facilities.

TOURIST IMAGES OF SALISBURY

An image survey of Salisbury was carried out by the Southern Tourist Board in 1995. Visitors were asked what they particularly liked about Salisbury, and also what they disliked (first five only are shown):

Positive images	Percentage
Countryside/rural landscape	28
Cathedral	16
Historical associations	11
Architecture	7
Character	6

Negative images	Percentage
Traffic congestion	12
Parking	10
Poor signposting	3
Building works	2
Cleanliness/litter	2

Visitors clearly see Salisbury very much as it was described at the beginning of this section, and regard vehicular traffic and parking as the two main tourism management issued.

STUDENT   10.3   ACTIVITY

1 Traffic congestion and inadequate car parking are the two main tourism management issues in Salisbury.
  Using the map of central Salisbury, what other recommendations could you make for the future, apart from increasing car parking capacity?

2 Salisbury does not possess any four or five star hotels and hotel bed occupancy in the city was 43 per cent in 1994.
  i Assess the merits of a site for a new hotel at Petersfinger, as proposed in the local plan (see Figure 10.4.4).
  ii What policy guidelines could be established for increasing bedspace occupancy in the city?

3 Construct a flow diagram showing the main tourist management issues facing historic cities such as Salisbury.

'Our objective, . . . must be to manage change in a way which recognises the national importance of such an historic landscape, the interest of all who own a part of it — and the opportunities as well as the constraints which such a heritage implies.'

Hadrian's Wall: World Heritage Site – Management Plan 1996

10.5 Managing an Ancient Site: Hadrian's Wall

Hadrian's Wall extends for nearly 185 km from Wallsend on the Tyne Estuary to Bowness-on-Solway (see map, Figure 10.5.1). It is the best preserved defence work in the former Roman Empire (Figure 10.5.2), and now presents a major management challenge to English Heritage. The Wall was designated as a World Heritage site in 1987, fulfilling the following criteria:

- it is an outstanding example of traditional human settlement and land use which is representative of a culture;
- it provides evidence of a civilisation that has disappeared;
- it illustrates a significant historical period.

The Wall traverses a varied set of

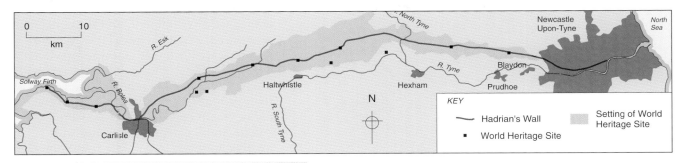

Figure 10.5.1   Hadrian's Wall and its World Heritage setting

landscapes. In the east, on Tyneside its environment is entirely urban: it passes through arable farmland in the east of Northumberland, whilst in the centre its surrounding landscape is more open and exposed, before returning to a more gentle landscape to the east of Carlisle. Beyond Carlisle in the west it runs along the edge of the tidal marshes of the Solway. The World Heritage site includes the essential linear elements of the Roman Frontier, with a series of outlying sites, such as

Figure 10.5.2   Hadrian's Wall, west of Housesteads

Corbridge Roman site, and the Vindolanda fort. It is enclosed by a buffer zone which should effectively cushion it from any inappropriate development.

The management of Hadrian's Wall involves a number of complex issues (Figure 10.5.3). Four major themes emerge:
- the conservation of the archaeological sites and their landscape;
- the maintenance of a well-established agricultural fabric within which the sites are set;
- controlled access to the sites for visitors;
- the contribution of the sites to the regional and national economy, through its tourist potential and related services.

Management of Hadrian's Wall has to reconcile the twin roles of conservation and tourism. Conservation presents particular difficulties in the urban areas of Tyneside and Carlisle (although these were excluded from the 1987 Designation). Nevertheless, the aim in these areas is to expose the extent of the remains wherever possible, and to create open space along its line for its fuller appreciation. In the rural

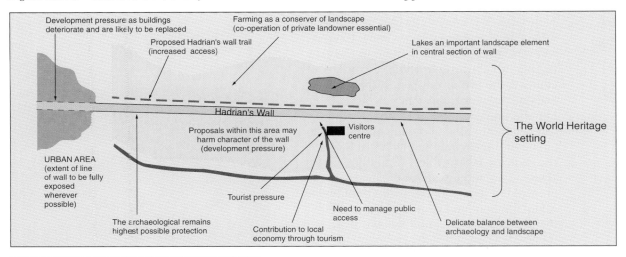

Figure 10.5.3   Management issues, Hadrian's Wall

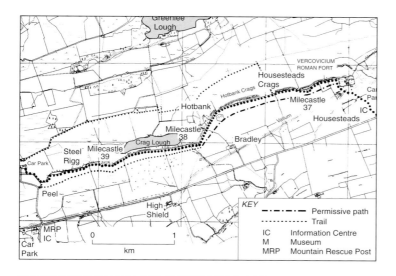

Figure 10.5.4   Trail from Steel Rigg to Housesteads

of tourism seeks to promote it in a controlled environment, and the concept of sensitivity (see Chapter Five, page 94) recognises three levels of tourist pressure. The monitoring of agreed environmental standards for each site, and the provision of relevant interpretive information will aid management of such tourist pressure. Access arrangements to the sites will involve an integrated public transport policy. A new National Trail (see section 10.7), the Hadrian's Wall Path, has been approved by the Department of the Environment, which will provide new opportunities for the management of walkers and the integration of rights of way.

STUDENT ACTIVITY

Study the Ordnance Survey extract of the section of Hadrian's Wall from Housesteads to Steel Rigg (Figure 10.5.4).
1 What are the essential features of the Wall and its buffer zone in this section?
2 a What would you consider to be the main management issues in this section?
  b How have they been addressed?

areas, Hadrian's Wall passes through a number of different local authorities, and most include specific schemes for its conservation within their Structure Plans. Co-operation of farmers and landowners is essential not only in the management of the archaeological sites, but also in creating an appropriate landscape setting for the World Heritage Site. Management

10.6 Managing New Lanark: A World Heritage Village

'Throughout its life, New Lanark has been a living, working community.'

New Lanark (Figure 10.6.1) is an industrial village in the upper Clyde valley. Cotton spinning mills were built there in the late eighteenth century. In the early nineteenth century, the philanthropist, Robert Owen became a managing partner at New Lanark, and set about introducing some fundamental social and educational reforms aimed at improving the quality of life of the growing labour force. Child labour was phased out, and Owen established some progressive styles of education, founding an infants' school and evening classes. With its free medical care, and its own store supplying goods cheaply, New Lanark was seen as a model industrial village in the nineteenth century.

Figure 10.6.1   New Lanark, Scotland

Figure 10.6.2   *Historical buildings, New Lanark*

The cotton mills closed in 1968, and since that time, conservation of the industrial village has been the main management theme at New Lanark. In the early 1970s New Lanark was designated as an Outstanding Conservation Area, and all buildings in New Lanark were classified as historic buildings Grade A. The New Lanark Conservation Trust was formed in 1974 to manage the restoration of the buildings, and to maintain New Lanark as a living, working community.

New Lanark is now a major tourist attraction in the Clyde Valley (Figure 10.6.3). A state-of-the-art visitor centre is the focus of tourist activity. In addition to the jobs created through the tourist facilities, about 20 small firms operate in restored buildings, with a hotel and conference facility planned for the late 1990s.

New Lanark Visitor Centre

Where and how to find us...

Lanark is a pleasant half hour's drive from the M74 at Abington, or less than an hour from either Glasgow, Stirling or Edinburgh. The village of New Lanark is signposted on all major routes. Rail access is via the station at Lanark. There are ample car parking facilities.

Details correct at time of going to print.

PRINTED BY MACDONALD LINDSAY PINDAR, LOANHEAD.

New Lanark Youth Hostel

Step back in time, stay in our 200 year old building in the heart of this famous conservation village. Travel back in time on the Annie McLeod Experience, see the 'stairheid cludgie', explore the Falls of Clyde Nature Reserve, visit the wildlife centre, the woollen mill, the classic car collection and much more!

For those who wish to cook for themselves the self-catering kitchen has everything you could need, cooking utensils, crockery are supplied - all you need to bring is the food! There's a comfortable dining room, a TV lounge, a laundry and a small shop.

Out of doors there's plenty to do. Step into the past - on the Annie McLeod Experience, a dark ride which takes you back in time to the sights and smells of 1820. Discover life in a bygone era at the Millworkers' House and the Village Store.

The walkways along the Falls of Clyde Nature Reserve let you enjoy the tranquillity and natural beauty of the area.

Figure 10.6.3   *New Lanark as a tourist attraction*

STUDENT ⬤10.5 ACTIVITY

1 Why has industrial archaeology become important tourist business in Britain?
2 What management difficulties would have been experienced in setting up the 'living and working community' at New Lanark.

Managing Britain's Longest National Trail: the South West Coast Path

10.7

These words were written in the middle of the nineteenth century, but still apply to the pleasures to be derived from long distance walking. National Trails, as they became known in 1988, are established 'to provide a nationally and internationally recognised series of walks and rides through the finest and most characteristic English landscapes, allowing the highest quality of experience and opportunities for both short and extended journeys within the same scenic corridor.'

The South West Coast Path is the longest of the National Trails of England and Wales (911 km – see Figure 10.7.2). Designation by the Countryside

'at last (I) emerged upon a headland, where you can settle in a nook of the rocks . . . then you can consume your modest sandwiches, . . . and feel more virtuous and thoroughly at peace with the universe . . .'

In Praise of Walking: Sir Leslie Stephen

Figure 10.7.1   *South West Coast Path, Crackington Haven, Cornwall*

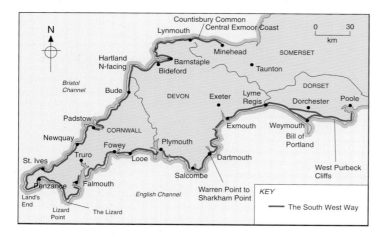

Figure 10.7.2   The South West Way

establishing an overall management strategy. In 1994, the Countryside Commission set up the South West Coast Path Project, which had two principal aims:

- a survey of the use and condition of the Path
- the production of an agreed strategy for the Trail

User Survey May–September, 1994

User profile
93% of the users were walking for one day or less
30% were walking for less than one hour
5% of the LDW were walking the whole Trail
56% of the LDW were on a walking holiday of 4–14 days

Origin
64% of the SDW were on holiday, and nearly one half lived more than 100 miles from the part of the Trail that they were using.
65% of the LDW lived more than 160 km away. Nearly 25% of the SDW lived within 16 km of the Trail (15% for LDW).

Occupation
More than half all respondents were in full time work (57% SDW, 61% LDW). Of SDW, about 75% were in professional, managerial or clerical employment. 18% of SDW were retired, and 14% of LDW.

Group structure
19% of SDW and 34% of LDW were walking alone. Only 2% were part of an organised group.
20% of SDW had one or more children under 15 with them, but only 5% of LDW.

(SDW = Short distance walker
LDW = long distance walker)

Commission occurred in a series of stages and it was opened in four distinct sections:

1973 The Cornwall Coast Path
1974 The South Devon Coast Path
1974 The Dorset Coast Path
1978 The Somerset and North Devon Coast Path

Integrated management of the South West Coast Path has proved to be a difficult task because of the many different authorities and bodies involved. Figure 10.7.3 illustrates the complexities that exist in

Type of use	Estimated number of user days
All of the Trail on this trip	5059
Part of the Trail on this trip	15 428
All of the Trail on separate trips	21 479
Short journeys (<one hour)	314 799
Part day journey (1–4 hours)	558 318
Full day journey (>4 hours)	158 948
Total	1 074 094

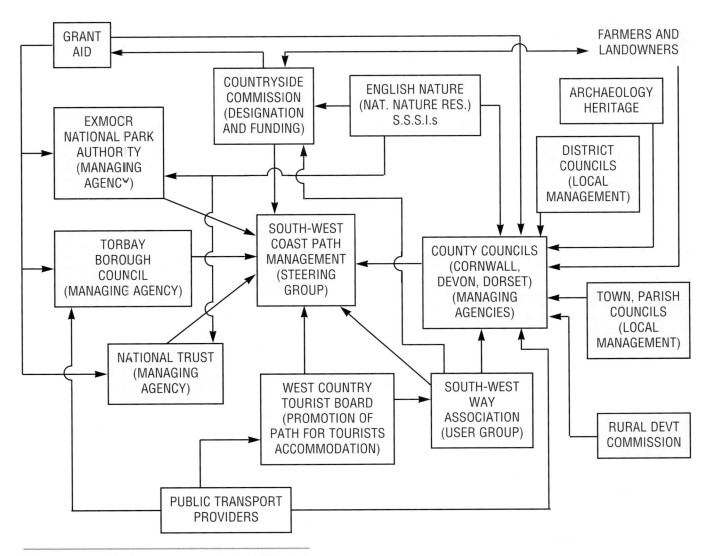

Figure 10.7.3   *Management structure, the South West Path*

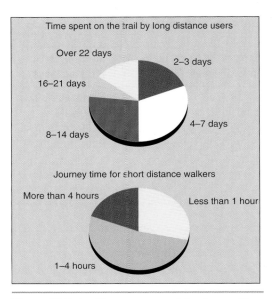

Figure 10.7.4   *Time spent on path by short and long distance walkers*

The visit
75% of the SDW knew they were walking a National Trail.
50% of the SDW knew their way along the trail, 25% were using waymarkers. Nearly all LDW were using maps of some kind. Nearly all walkers thought that waymarking was sufficient, except through towns and villages.

Transport
Most SDW used private transport to reach the Trail (59%), although 39% arrived on foot. Only 35% used public transport, but 53% said that they would use a bus if it were available.
The LDW did not rely on private transport as much (50%), and 35% used public transport to reach the Trail, and 40% on their return.

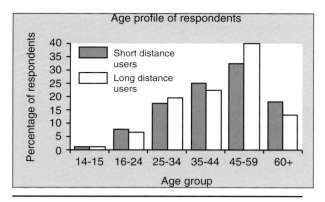

Figure 10.7.5 *Age structure of users of the South West Path*

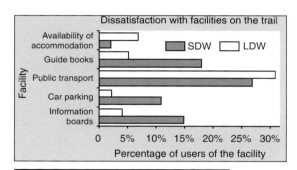

Figure 10.7.6 *Rating of facilities on the trail*

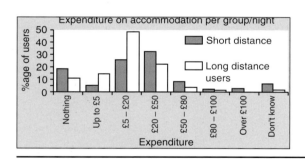

Figure 10.7.7 *Expenditure on accommodation per group/night*

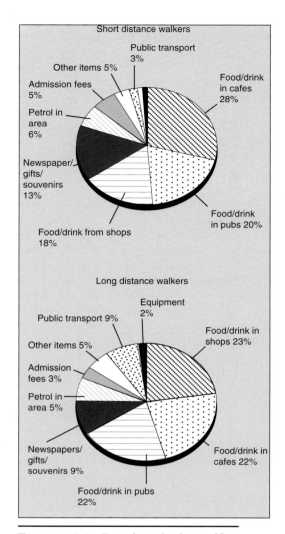

Figure 10.7.8 *Expenditure for short and long distance walkers on the trail*

Attitudes
Scenery was the main feature that attracted people to the path (63% SDW and 93% LDW).

STUDENT ⬤10.6 ACTIVITY

1 Discuss the use that can be made of the user survey (pages 180–2) in management planning of the Trail.
2 What other information, not mentioned above would be useful to collect from users of the path.

Condition of path

Natural path surfaces path surfaces	Man-made Path surfaces
1 <33% vegetation loss	First class
2 33–66% vegetation loss	Satisfactory
3 >66% vegetation loss	Deteriorating
4 5–10 cm soil loss	Poor requires attention
5 >10 cm soil loss	Replacement required

STUDENT ⬤10.7 ACTIVITY

Five locations are shown on the map of the South West Coast Path (Figure 10.7.2). Details of the path condition along each stretch are given below.

Location	Condition of path
Central Exmoor Coast	4% vegetation, 91% track, 5% urban, all classes 1–3
Hartland – north facing	67% vegetation, classes 1–5 (19% in 4 and 5) 25% track, 2% urban – classes 1–3
The Lizard	81% vegetation, 3% sand and shingle, 2% rock, classes 1–4 (45% in 4) 5% track, 8% urban, 1–3
Warren Point to Sharkham Point	100% vegetation, classes 1–3
West Purbeck cliffs	69% vegetation, 3% rock – classes 1–4 (20% in 4), 22% track, 6% urban, classes 1–2

Location pedestrian counts	
Central Exmoor Coast	5373
Hartland – north facing	3262
The Lizard	49 570
Warren Point	n/a
West Purbeck	120 527

1 Using the map (Figure 10.7.2) and the details for the five locations, explain the variations in path condition discovered in the five locations.
2 What management action would be required in the five stretches of coastline surveyed?

10.8 Managing an Area of Outstanding Natural Beauty: the Malvern Conservators

The Malvern Hills are one of England's most striking landscapes, rising in unmistakeable profile from the Severn lowlands to the east, and brooding over the rich farmscapes to the west (Figure 10.8.1). The AONB includes not only the Malvern Hills but other attractive rural tracts of country to the north, south and west: it occupies 105 km², and is the eighth smallest of the 35 designated AONBs in England.

The ridge of the Malvern Hills, built principally of Precambrian rocks, extends 13 km from north to south (Figure 10.8.2). It possesses a number of distinctive summits that together create what Stanley Baldwin eulogised as 'the most beautiful silhouette in the world'. Much of the ridge carries a cover of acid grassland, a product of centuries of sheep grazing. Lower down the slopes, oak-hazel woodland provides a contrast, particularly effective in autumn. Human influence contributes strongly to the character of the Malvern Hills. On the east distinctive Victorian buildings, often painted white, extend up the slopes from

Great Malvern, whilst on the west a gentler note is struck in the mellow brick and timber villages at the foot of the Hills. Although management guidelines exist by virtue of their AONB status, day-to-day management of the Malvern Hills is carried out by the Malvern Conservators. The Board of Conservators was incorporated under the Malvern Hills Acts 1884 to 1995. The Conservators are responsible for most of the Hills, and the

'If you are walking on the hills and hear music, don't be afraid, it's only me.'

Sir Edward Elgar

Figure 10.8.1 *The Malvern Hills*

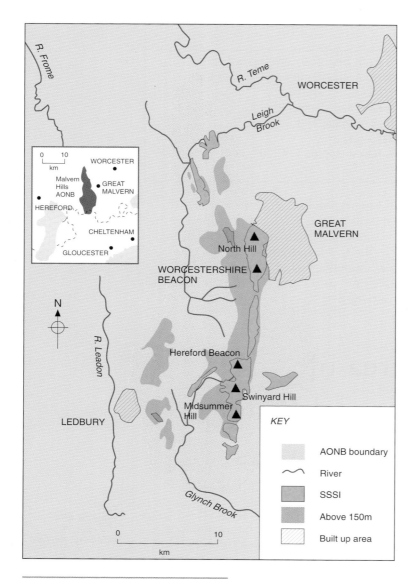

Figure 10.8.2   *The Malvern Hills AONB*

Figure 10.8.3a   *Worcester Beacon, Malvern Hills*

Figure 10.8.3b   *Tracks on the Malvern Hills*

Figure 10.8.3c   *Wynds Point car park, Malvern Hills*

Figure 10.8.3d   *Hollybed Common, Malvern Hills*

surrounding Commons, together with some other stretches of land to the east and west. The Malvern Hills Acts require the Conservators to manage the Hills under certain statutory obligations:

- to prevent encroachments on the Hills;
- to keep the Hills open and unbuilt on for the benefit, recreation and enjoyment of the public;
- to conserve and enhance the character and quality of the existing landscape;
- to conserve and enhance the existing wildlife of the Hills and Commons;
- to protect the interests of Commoners.

Four further aspects of management of the Hills are considered important enough for the Conservators to include in their

Management Plan:
- to provide opportunities for informal outdoor recreation;
- to improve the public's knowledge, understanding and respect for the hills;
- to have regard for the social and economic well-being of the people living within the area;
- to seek to influence planning control and development in the area.

STUDENT 10.8 ACTIVITY

1 Draw up a compatibility matrix to show that some of these objectives might be in conflict, with one another. How might such conflict be resolved within a management plan?
2 The Conservators are required to manage both the natural and man-made resources of the Hills. Using the photographs in Figure 10.8.3 and other information in this section suggest what general management guidelines might be appropriate for
 a low lying commons adjacent to the Hills
 b high open hill areas
 c abandoned quarries (the last quarry closed in 1977)
 d footpaths and tracks
 e car parks and parking areas.

ESSAYS

1 Better management of tourist resources could reduce much tourist pressure. How far do you agree with this statement?
2 Tourist pressures in the countryside are different to those in the town. Discuss this statement.

FIELDWORK OPPORTUNITIES & PROJECT SUGGESTIONS

1 Select any small or medium sized park. Attempt to discover whether a management plan exists or not. If the plan is available, use it as a basis for determining the main forms of management used in the park. It might be useful to compare the forms of management with those suggested in the Bromley report in section 10.3.
2 Measure the environmental impact of heavy visitor use on a popular tourist site. Visitor density and visitor activity can be plotted. Level of use of car parking facilities can be surveyed or informal car parking can be measured. Environmental impact such as damage to vegetation by people and vehicles can also be plotted. An assessment of the effectiveness of management initiatives at the site can be used for a summary.

Glossary

ADAS: (Agricultural Development and Advisory Service) Government organisation, operating within MAFF (Ministry of Agriculture, Fisheries and Food)

algal bloom: prolific growth of algae on a body of water, usually due to eutrophication

berm: ridge of shingle on a beach

biofiltration: use of plant material, e.g. peat, as a filter for noxious substances

brown forest soils: soils that develop under a deciduous forest cover in middle latitudes

bryophytes: plant group to which liverworts and mosses belong

carrying capacity: the maximum stocking level of a pasture

cauldron subsidence: volcanic activity associated with a ring fault

clear felling: felling of total tree population in one area

climatic climax: the plant community that forms naturally under particular climate conditions

commoner: people who have grazing and other rights on common land

commoning: the exercising of the above rights

convenience goods: everyday goods such as food, that are purchased frequently

coppicing: the practice of cutting trees to their stumps, and allowing regeneration from the stumps

counterurbanisation: the process of population moving from urban areas into rural areas

cutaway bog: bog that has been exploited for peat

dBA: noise measurement level

de-industrialisation: the decline of manufacturing industry

deregulation: the freeing of bus services from controls to stimulate competition

dolerite: a tough resistant basic igneous rock

durable goods: goods which are usually an occasional purchase, such as furniture

ecological succession: the change in an ecosystem whereby one group of organisms is replaced by others over a period of time

ERDF: European Regional Development Fund

eutrophication: addition of nutrients to a water environment

fluvioglacial deposition: the deposition of material by meltwater from a glacier or an ice sheet

GDP: (Gross Domestic Product) the total amount of goods and services produced by a national economy in a given time

gentrification: the occupation of a former working class area by middle class groups

geomorphology: the study of landforms

headland: area of unploughed land around the edge of fields

hierarchy (of open spaces): a system within which open spaces are ranked in a clearly defined order to show their status

holistic: an approach where the whole of a system is considered more important than its parts

injection wells: wells in an oilfield down which mud or water may be pumped to assist drilling and recovery of oil

intangible (benefits): those benefits (in a cost benefit analysis) which cannot be assigned a monetary value

isotatic (subsidence): subsidence of the earth's crust as a response to overlying mass, e.g. an ice sheet

just-in-time: the delivery of components at the precise moment when they are required in manufacture

karst: scenery that is developed on well-bedded and permeable limestone

leachable components: soil components that can be washed downwards in the soil profile under acid conditions

Liassic: the lowest (oldest) division of the Jurassic system

lichen test: tests using frequency of occurrence and well-being of lichens to determine the level of atmospheric pollution

lithology: the study of rock characteristics

machair: highly calcareous shelly sand that fringes many of the windward Hebridean coasts, often covered by a rich grassland ecosystem

marine transgression: spread of the sea over a neighbouring land mass

Ml/day: megalitres per day

multiplier effect: additional changes to an economy consequent upon an initial one

myxamatosis: fatal viral disease of rabbits

net drift: the final amount of movement of

beach material over a given period of time

neutral grasslands: grasslands that are growing in a soil environment that is neither acid or alkaline

nutrient stripping: the removal of nutrients from a water body

perambulation: the boundary of a body of land, e.g. the New Forest

plagioclimax: a stable plant community resulting from the action of people on a climax community

planning blight: the adverse effects of long term planning proposals

Pleistocene Period: the Ice Age period of geological time

podzolic: term used to refer to podsols, soils that are extremely acid, and usually infertile

pollarding: the practice of cutting off the branches of a tree to encourage young growth

polygonal joints: joints in igneous rocks that have developed during cooling

quango: semi-public body with financial support from the Government

Ramsar site: protected wetland conservation site

return period: the average length of time separating events (e.g. floods) of similar magnitude

right-to-buy: the right of council house tenants to purchase the house in which they live

ring dyke: igneous body that has been injected along a circular fault

rip current: turbulent body of sea water where tidal currents meet or wind generated waves meet tidal currents

riparian: located on the banks of a river

rural deprivation: standard of living below the norm in rural areas

schist: medium grained metamorphic rock

severance: the division of a property, e.g. a farm by the route of a bypass

sieving exercise: map exercise in which various data elements are progressively removed (sieved out) in order to clarify distributions

SSSI: Site of Special Scientific Interest

sunrise industries: new industries, usually involving high technology, with prospects of rapid growth and development

tangible: benefits to which a monetary value can be assigned in a cost/benefit analysis

tear fault: fracture in rocks where the sense of movement is horizontal rather than vertical

tertiary activities: service industries

Tertiary: era of geological time which began approximately 65 million years B.P.

transnational: industrial companies that have plants in a number of different countries

TTWA: Travel to Work Area

volcanic vents: the opening through which volcanic material is ejected

wave climate: direction and strength of waves at any one location

Index